José Victor Ferreira Fernandes
Kristerson Reinaldo de Luna Freire
Adna Cristina Barbosa de Sousa

PHYSICOCHEMICAL CHARACTERIZATION OF PLEUROTUS OSTREATUS VAR. FLORIDA

José Victor Ferreira Fernandes
Kristerson Reinaldo de Luna Freire
Adna Cristina Barbosa de Sousa

PHYSICOCHEMICAL CHARACTERIZATION OF PLEUROTUS OSTREATUS VAR. FLORIDA

Characterization of white Shimeje grown on co-products from bean cultivation and beer production

ScienciaScripts

Imprint

Cover image: www.ingimage.com

This book is a translation from the original published under ISBN 978-613-9-72691-2.

Publisher:
Sciencia Scripts
is a trademark of
Dodo Books Indian Ocean Ltd. and OmniScriptum S.R.L publishing group

120 High Road, East Finchley, London, N2 9ED, United Kingdom
Str. Armeneasca 28/1, office 1, Chisinau MD-2012, Republic of Moldova, Europe
Printed at: see last page
ISBN: 978-620-8-35386-5

PHYSICO-CHEMICAL CHARACTERISATION OF PLEUROTUS OSTREATUS VAR. FLORIDA

CHARACTERISATION OF WHITE SHIMEJE GROWN ON CO-PRODUCTS FROM BEAN CULTIVATION AND BEER PRODUCTION

JOSÉ VICTOR FERREIRA FERNANDES[1]

FEDERAL UNIVERSITY OF PARAÍBA CAMPUS I/BIOTECHNOLOGY CENTRE - CBIOTEC)

DR. KRISTERSON REINALDO DE LUNA FREIRE [2]

LECTURER AT THE FEDERAL UNIVERSITY OF PARAÍBA CAMPUS I/CENTRE OF BIOTECHNOLOGY - CBIOTEC

DR[A] . ADNA CRISTINA BARBOSA DE SOUSA[3]

LECTURER AT THE FEDERAL UNIVERSITY OF PARAÍBA CAMPUS I/CENTRE OF BIOTECHNOLOGY - CBIOTEC E-MAIL CONTACT: ADNACSOUSA@GMAIL.COM

ACKNOWLEDGEMENTS

Firstly, I would like to express my deep gratitude to my parents, Ananelis Fernandes and José Nildo Ferreira. Without their unconditional support, I would never have got this far and this text would not exist. Another fundamental pillar in my development is, and always will be, my brother, Vinicius, to whom I owe my sincere thanks for always encouraging me to improve and surpass him. Still with regard to my family, I extend my gratitude to my grandparents, especially my grandmother Francisca and grandfather Sebastião, as well as my cousins, uncles and aunts, especially those from João Pessoa, who helped me in different ways. Although there is not enough space to mention each one in detail, they all occupy a special place in my memory.

I dedicate this paragraph as a token of my sincere gratitude to my girlfriend, Hellen Sousa. She has undoubtedly been essential to my personal, social and professional growth. I thank her not only for sharing in my achievements, but above all for being there for me in difficult times, always encouraging me to improve. I am immensely grateful for her dedication, affection and for being by my side at every stage of this journey.

I would like to thank my dear friends. Distance will never be able to break the connection we have created. Thank you to Davi Mendes and his family who, when I arrived in João Pessoa with no resources, helped me financially without asking for anything in return. I'll never forget what they did for me. Davi, my oldest friend, is my first reference of true friendship; I can't express all my gratitude in words, but I try to convey it in every hug we give. I would also like to thank Garrido, a childhood friend for whom I have deep admiration, who was able to test my self-control and make me realise how patient and moderate I am. To João Pedro, I am grateful for his constant presence and for all the help he offered me during my degree. My sincere gratitude also goes to Nattanael, who was the second most present person in my life during the years I spent in João Pessoa. He took me in when I had nowhere to stay, was deeply sincere in his thoughts and attitudes, and helped me in countless ways over the years. To all the other important people who have not been mentioned here, know that you are kept in a special place in my memories.

I would like to thank the Federal University of Paraíba (UFPB). This institution provided not only a solid foundation of knowledge, but also a welcoming and challenging environment. My gratitude goes to the Biotechnology Centre, along with all its staff and technicians, for providing the physical structure needed to carry out the work. As well as the Molecular Genetics and Plant Biotechnology Laboratory (LGMBiotec/UFPB) and the Food Technology Laboratory (LTA/UFPB). I would also like to thank the CNPq (National Council for Scientific and Technological Development) for the student funding, without which it would not have been possible to carry out this research

Finally, I would like to express my deep gratitude to my supervisor, teacher and, above all, friend Adna Cristina. During the years I have been under her guidance, I have learnt much more than academic content; Professor Adna has passed on valuable lessons for life, guiding me to this moment. It's only right to end with my sincere thanks to her.

SUMMARY

Edible mushrooms of the Pleurotus genus are widely recognised for their ecological, biotechnological and nutritional applications, and are the second most cultivated type of mushroom in the world. This study investigated the production of Pleurotus ostreatus var. florida grown in substrates based on agro-industrial by-products from bean cultivation and craft beer production. Four formulations were tested, varying only the type of supplement used: T1 (80 per cent green bean pods without grains + 20 per cent malt bagasse), T2 (T1 + 5 per cent residual yeast), T3 (T1 + 10 per cent hot trub) and T4 (T1 + 10 per cent hop trub). The substrates and supplements were analysed for pH, reducing, non-reducing and total sugars, ash, soluble solids, carbon content (C), nitrogen content (N) and C/N ratio. The mushrooms were analysed for biological and production parameters, including moisture, protein, lipid, carbohydrate, ash and total calorie content. The results showed that the formulation with hop trub (T4) had the highest yield (13%), biological efficiency (61%) and productivity (1.54 g.day^{-1}), while the fastest production cycle (22 days) was observed in the formulations with hot trub (T3) and hop trub (T4). In nutritional terms, T3 stood out with the highest protein content (28.82 per cent), the lowest carbohydrate value (56.48 per cent) and a calorific value of 422.02 Kcal/100g. The different formulations showed promising potential for producing mushrooms with high nutritional value, especially the formulation with hot trub supplementation (T3), which stood out for its nutritional quality, and cold trub (T4) due to its higher yield. Future research could explore the optimisation of these substrates and the evaluation of new agro-industrial waste, contributing to sustainability and innovation in fungiculture.

Keywords: *Edible mushroom, agro-industrial co-products, fungiculture, Pleurotus spp.*

SUMMARY

1. INTRODUCTION

Pleurotus ostreatus, popularly known as the oyster mushroom, is a species of edible fungus that is widely cultivated and valued for its nutritional properties and health benefits (El-Ramady et al., 2022). The growing demand for sustainable and nutritious food has boosted interest in mushrooms, such as P. ostreatus, in both the national and global markets. Rich in protein, essential amino acids, omega-6 and fibre, this mushroom has attracted the attention of the food and pharmaceutical industries due to its bioactive compounds with antioxidant, antimicrobial and immunomodulatory properties. With the global mushroom market projected to grow at an annual rate of 9.7 per cent until 2030, fungiculture stands out as an activity with great potential (Grand View Research, 2021). In Brazil, although mushroom consumption is still relatively low compared to other countries, production has been growing significantly, signalling a promising future for P. ostreatus in the local market.

Axenic mushroom cultivation, which is characterised by the development of the fungus in sterilised organic substrates, has been shown to be effective in reducing costs and improving production by replicating natural environmental conditions (An et al., 2021). This method not only maximises production, but also promotes sustainability by recycling agro-industrial by-products, inserting the process into a circular economy (Vaz Junior, 2020). By using agricultural and industrial by-products, such as rice straw, sugar cane bagasse and corn by-products, Pleurotus cultivation contributes to reducing pollution and conserving natural resources (Raman et al., 2021).

Among the most promising co-products for the production of P. ostreatus are waste from beer production, such as malt pomace, hot and cold trub, and residual yeast, which are often not properly reused (Amoriello & Ciccoritti, 2021). In addition, bean pods, an abundant by-product in Brazil, are rich in nutritional compounds such as organic acids, phenolics, vitamins, proteins and fibres, making them a suitable substrate for bioconversion into edible protein by fungi (Omura et al., 2020). The aim of this study was to evaluate the production and nutritional aspects of P. ostreatus var. florida basidiomata grown in different substrate formulations using by-products from bean cultivation and craft beer production.

2. OBJECTIVES

2.1. OBJECTIVE GENERAL

Evaluate the production and nutritional aspects of the basidiomata of P. ostreatus var.florida grown in different formulations.

2.2. OBJECTIVES SPECIFIC

To evaluate the potential for using bean pods without the grain, supplemented with malt bagasse, hot trub, cold trub and residual yeast in the formulation of a growing medium for mushroom production;

Characterise the individual substrates in terms of pH, reducing sugars, total reducing sugars, ash and total soluble solids;

Characterise the biological parameters: substrate colonisation time, primordia emission period, basidiom formation time, basidiom diameter, pileus height and pileus width;

Characterise the production parameters: yield, biological efficiency, productivity, compost consumption and loss of organic matter of the different formulations;

Physico-chemical characterisation of the mushrooms in terms of total carbohydrates, total proteins, lipids, ash, moisture and total calorific value.

3. THEORETICAL FOUNDATION

3.1. Aspects biological

The Pleurotus genus comprises a great diversity of species, including P. florida, P. ostreatus, P. eryngii, P. pulmonarius, P. sajor-caju and their varieties, all of which are considered edible mushrooms (Jongman et al., 2018). These species stand out for their rapid growth compared to other mushrooms, and their fruiting bodies are rarely affected by pests or diseases. The cultivation of these species is simple, economical and high-yielding, and they can be grown in a variety of substrates, with a wide tolerance to different chemical and temperature conditions (Bellettini et al., 2019). Among the more than 40 known species in this genus, P. ostreatus var. florida stands out not only for its sensory characteristics of odour, taste and nutritional value, but also for its biotechnological and medicinal applications (El-Ramady et al., 2022).
Taxonomically, P. ostreatus belongs to the Kingdom Fungi, Phylum Basidiomycota, Class Agaricomycetes, Order Agaricales, Family Pleurotaceae and Genus Pleurotus (Toros et al., 2022). It is popularly known as Shimeji, Hiratake, oyster mushroom, due to its shape resembling an oyster when cultivated, or white rot fungus because it occurs spontaneously on decaying tree trunks (Figure 01).

Figure 1. Pleurotus ostreatus colonising a tree trunk

Source: Albuquerque et al (2024)

In nature, several species of this genus can be found in tropical forests, gardens, backyards, around rice fields, on dead wood trunks and on decaying organic matter (Daud et al., 2021). They often grow on decaying tree trunks, fallen branches and damp trunks (Barh et al., 2019), especially on trees such as carpinus (Carpinus sp.), beech (Fagus sp.), willow (Salix sp.), poplar (Populus sp.), birch (Betula sp.) and walnut (Juglans regia) (El-Ramady et al., 2022).
In its vegetative form, P. ostreatus var. florida consists of a cluster of undifferentiated hyphae, forming a dense mycelium that spreads across the substrate. These hyphae can differentiate and form fruiting bodies in response to specific environmental changes, such as variations in temperature, CO levels$_2$, O availability$_2$ and light intensity (Sydor et al., 2022). These

environmental conditions are crucial for initiating the process of differentiation and basidiome formation.Morphologically, P. ostreatus var. florida is recognisable by its smooth, flattened, irregularly-shaped hat, similar to a fan or oyster. The hats are convex when young, flattening out and increasing in size as the basidium ages. The edges of the hat are also irregular and change over time. The lamellae, which make up the hymenium, occupy the space between the edges of the hat and the stipe. The colour can vary from white to grey or brown, depending on the variety and the stage of development of the mushroom. They generally grow in overlapping clusters and may not develop a stipe when growing laterally. In addition, P. ostreatus var. florida has a delicate and mild flavour, sometimes described as similar to shellfish, with a firm, fleshy texture and can be eaten in a variety of ways (Albuquerque; Sousa, 2024).

Figure 2: Morphological structure of Pleurotus osteatus var. florida

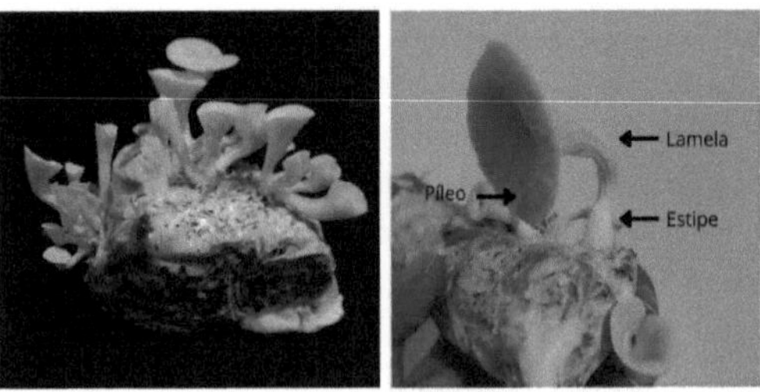

Source: Author (2024)

In terms of nutrition, they are heterotrophic by absorption, producing enzymes that facilitate the degradation of organic matter and, consequently, help in the absorption of micro and macro elements that are important for their growth. P. ostreatus uses glucose and fructose, derivatives of lignin, cellulose and hemicellulose commonly found in biodiversity and agricultural systems, as its main carbon sources (Sun et al., 2023).

Lignin degradation occurs through the joint action of several enzymes, some of which take part in various reactions involving veratryl alcohol (VA) (Figure 03). An important example of indirect participation in this process is the action of the glyoxal oxidase enzyme, which catalyses the reaction of glyoxal with oxygen, releasing hydrogen peroxide (H O_{22}) and glyoxylic acid. The enzymes lignin peroxidase and manganese peroxidase, which are directly involved in lignin degradation, need hydrogen peroxide as a cofactor to start their catalytic activity. Lacases, which have been widely documented in lignin degradation, are essential to the ligninolytic mechanism as they only require oxygen to catalyse the degradation of aromatic compounds (Durán-Aranguren et al., 2021).

Figure 3 Enzymatic mechanisms in the degradation of lignin by Pleurotus ostreatus var. florida.

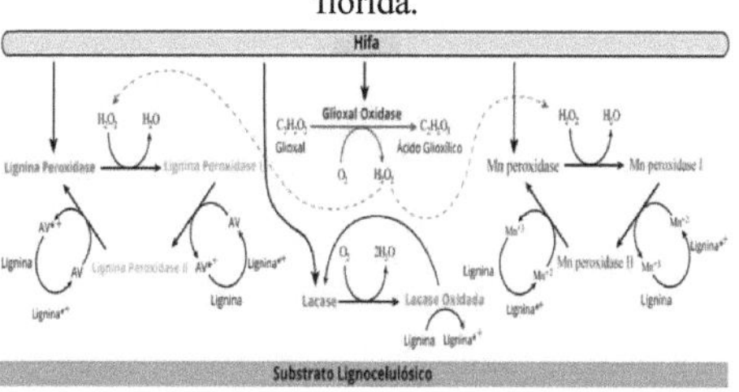

Source (Adapted from Durán-Aranguren et al., 2021)

Another important group of enzymes are hydrolases, which catalyse the depolymerisation of cellulose and hemicellulose. In the degradation of cellulose (Figure 04) into simpler sugars, three types of enzyme stand out: exoglucanases, endoglucanases and B-glucosidases. Exoglucanases cleave the B-1,4-glycosidic bonds in cellulose polymers, removing cellobiose (a glucose dimer) or glucose from the ends of the cellulose chain, the crystalline region where the fibres are densely packed and highly ordered. Endoglucanases act inside the cellulose chains, breaking the B-1,4-glycosidic bonds randomly along the chain, resulting in the formation of oligosaccharides of varying lengths and creating new endpoints that can subsequently be attacked by exoglucanases. Endoglucanases are most effective in the amorphous regions of cellulose, where the fibres are less ordered and more accessible. B-glucosidases play a crucial role in the final stage of cellulose degradation, hydrolysing cellobiose and other soluble oligosaccharides into glucose (Durán-Aranguren et al., 2021).

Figure 4: Action of hydrolase enzymes in the degradation of cellulose.

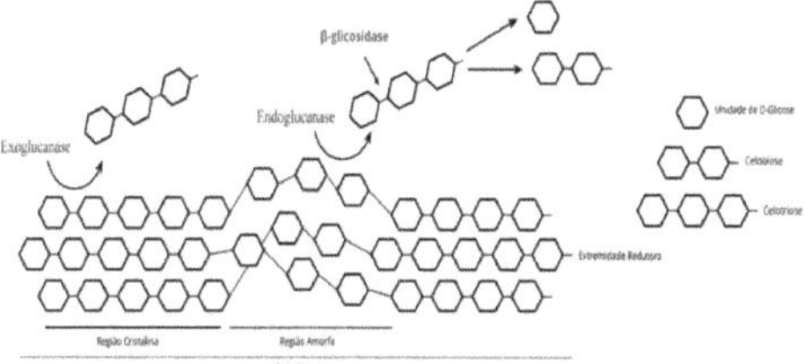

Source (Adapted from Durán-Aranguren et al., 2021)

Hemicellulose is a complex component of plant cell walls, made up of a heterogeneous mixture of polysaccharides. Unlike cellulose, which is composed exclusively of glucose units, hemicellulose has a more diverse structure, containing various pentoses (such as xylose and arabinose) and hexoses (such as galactose, mannose and glucose), as well as uronic acids and other substituents (Figure 05). Hemicellulose degradation involves the coordinated action of several specific enzymes, each responsible for cleaving different chemical bonds within the polymer structure (Zamora et al., 2021).

Figure 5: Action of enzymes in the degradation of hemicellulose.

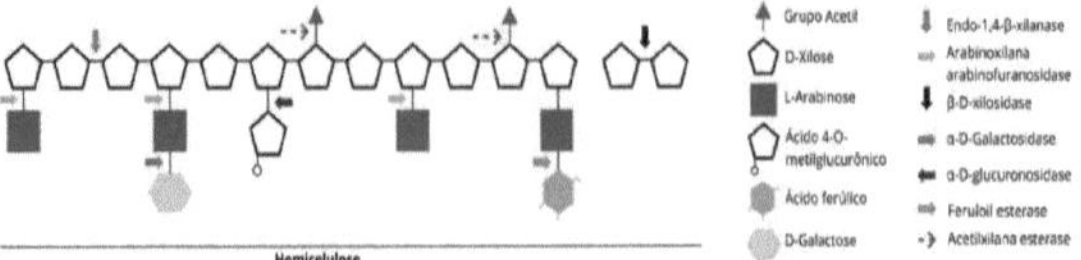

Source (Adapted from Durán-Aranguren et al., 2021)

From the degradation of organic matter by the action of different enzymes, the fungus completes its life cycle (Figure 06). The life cycle of P. ostreatus var. florida begins with the germination of haploid basidiospores, which develop into haploid mycelia. When two compatible haploid mycelia meet, fusion (anastomosis) occurs, forming a dikaryotic mycelium, which contains two unfused nuclei. The dikaryotic hyphae, in an appropriate culture medium, accumulate and form the basidiome, also called a mushroom (Figure 07A). The lamellae are found at the bottom of the basidiome, where the basidia are located. Within these differentiated structures at the hyphal ends, karyogamy occurs and then meiosis (Albuquerque; Sousa, 2024). Each basidium generates four basidiospores, sub-cylindrical, smooth sexual spores with haploid nuclei (Figure 07B). Upon germination, these spores transform into haploid mycelia, restarting the life cycle (Barh et al., 2019).

Figure 6. Life cycle of Pleurotus ostreatus var. florida.

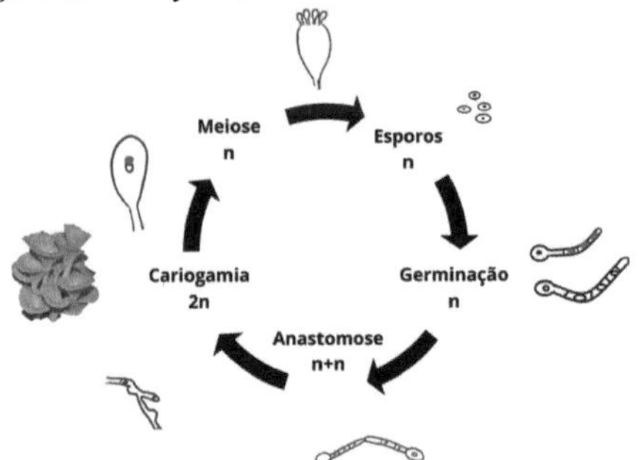

Source: Adapted from Barh et al. (2019).

Figure 07: Fruiting bodies and spores of Pleurotus ostreatus var. florida.

Figure 7. Fruiting bodies and spores of Pleurotus ostreatus var. florida.

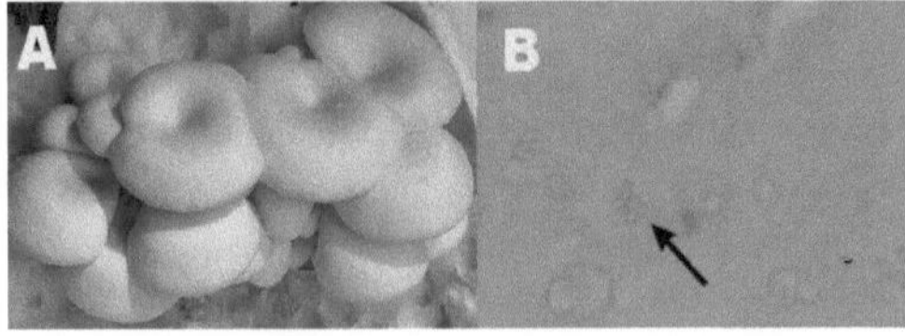

A: Upper part of the basidiome; **B**: Sexual spores (basidiospores). Source: Author, 2024.

3.2. Nutritional composition and bioactive compounds

Nutrients can be classified into two groups: micronutrients, which include vitamins and minerals, and macronutrients, which include carbohydrates, proteins and fats. Micronutrient requirements in humans are generally in quantities below 100 mg/day, in contrast to macronutrients, which are required in grams per day (Godswill et al., 2020). Furthermore, within nutrients, some are considered essential, i.e. they are necessary for human life and for tissue growth and repair. In contrast, bioactives are defined as constituents in food or dietary supplements, other than those required to meet basic human nutritional needs, which are responsible for changes in health status (Erdman, 2023).The ability of P. ostreatus to nutritionally enrich the substrates used as a result of its colonisation is widely reported in the literature. For example, Hussein et al. (2019) observed that, over a 28-day cultivation interval, there was an increase in the content of proteins (3.64%-9.16%), minerals (8.21%-8.64%) and lipids (2.4%-4.8%) in wheat colonised by the fungus. However, its application in the context of food is not limited to this. Pleurotus species are considered to be sources of carbohydrates, lipids, proteins, mineral salts, iron, important B vitamins and calcium, as well as having functional characteristics of different bioactive compounds (Toros et al., 2022). The fruiting body contains approximately 100 different bioactive compounds, including phenolic acids, flavonoids, hydroxycinnamic acids, hydroxybenzoic acids, lignans, tannins, stilbenes and oxidised polyphenols (Assemie; Abaya, 2022). The vegetative part (mycelium) of P. ostreatus is rich in beta-carotene, ascorbic acid, gallic acid, non-starch polysaccharides, beta-glucan, lovastatin, steroids and niacin.

These compounds have antioxidant, antidiabetic, antihypercholesterolaemic, anticancer and immunomodulatory properties (Atlh et al., 2019). The amount of macro and micronutrients varies according to the extrinsic and intrinsic factors involved in production (Table 01).

Table 1. Variation in the nutritional composition of Pleurotus ostreatus in different formulations.

Substratos residuais	N°	Proteinas(%)	Lipidios(%)	Cinzas(%)	Carboidratos(%)	Fibras(%)	Autores
Algodão	17	20.16 – 28.73	1,2 – 7,23	4,9–10,62	14,81–65,9	4,08 – 12,5	(Obiaigwe *et al*., 2023; Sardar *et al*., 2020). (Odunmbaku, 2018)
Palmeira de dênde	10	10,09-19,14	1,30 - 3,44	4,12-10,10	2,11-6,19	9,89-16,89	(Elkanah *et al*., 2022)
Arroz e trigo	7	19,21 – 29,76	3,12 – 5,64	5,89 – 7,87	36,68 – 50,96	10,09 - 13,38	(Aisha. *et al*., 2019)
Cana-de - Açucar	5	20,16-30,46	2,09-2,59	7,21-8,07	54,55- 58,50	4,08-5,14	(Obiaigwe *et al*., 2023)
Sisal	5	24,6 – 29,0	2,0 – 3,0	6,6 – 7,2	38,6 – 41,6	22,2 – 24,2	(Oliveira Do Carmo *et al*., 2021)
Milho	2	19,26-26,34	4,78-5,20	4,45-4,48	38,41-49,03	4,36-4,39	(Ewekeye *et al*., 2020)
Banana	1	15,03	1,31	7,99	26,41	5,88	(Garuba *et al*., 2017)
Mandioca	1	16,47	2,18	11,21	24,92	5,14	(Garuba *et al*., 2017)
TOTAL	48	10,09 – 30,46	1,2 – 7,23	3,74-11,21	2,11 – 65,9	4,08 –24,2	

NI = Não Informado. N° = Número de formulações. Fonte: autoria própria, 2024.

3.2.1. Macronutrients

Proteins: proteins of fungal origin have gained the attention of the food industry and the scientific community due to their high nutritional values, associated with the rich level of essential amino acids. The protein content of mushrooms is generally higher compared to other vegetables, but lower compared to meat and milk. In addition, they contain all the essential amino acids, including sulphur-containing amino acids such as methionine and cysteine (Toros et al., 2022). The amount of protein can vary between 10.09% and 30.46% (Table 01). Some essential and non-essential amino acids present in Pleurotus ostreatus mushrooms can be seen in Table 02.

Table 2. Amino acid profile of Pleurotus ostreatus.

Essential*/Non-Essential Amino Acids	Concentration (mg/100g)
Leucine*	135,670
Threonine*	10,4144
Methionine*	10,3644
Phenylalanine*	10,3028
Lysine*	231,824
Alanine	10,0342
Aspartic Acid	492,1208
Proline	10,4191
Serina	10,4585
Asparagine	10,4385
Hydroxyproline	9,3368
Cysteine	9,3234
Glutamine	12,0376

Source: (Adapted from Effiong et al. 2024)

Carbohydrates: mushrooms are recognised as an excellent source of carbohydrates in the form of polysaccharides, including glycogen, dietary fibres, cellulose, B-glucans, as well as some hemicelluloses such as galactans and xylans (Ewekeye et al., 2020). In addition, they contain a rich set of oligosaccharides, ergothioneine, monosaccharides and disaccharides (Table 03), which perform prebiotic functions and are highly beneficial for maintaining a healthy gut microbiota (Effiong et al., 2024). The amount of carbohydrates can vary between 2.11% and 65.9% (Table 01). They are considered a valuable source of edible dietary fibre due to the presence of non-starch polysaccharides, and provide essential food compounds that are valuable for human nutrition (Ewekeye et al., 2020). Dietary fibre in mushrooms is mainly composed of chitin, a straight-chain (1---4)-B-linked polymer of N-acetylglucosamine, and polysaccharides such as (1---3)-B-D-glucans and mannans present in their cell walls (Raman et al., 2021). The amount of fibre can vary between 4.08% and 24.2% (Table 01).

Table 3. Mono- and disaccharide profile of Pleurotus ostreatus.

Mono/Disaccharides*	Concentration (g/100g)
Erythrosis	0,4785
Glucose	55,0888
Xylose	7,1869
Fructose	19,6984
Galactose	17,466
Cellobiose*	0,0068
Trehalose*	7,3733
Chitobiose*	11,7858
Maltose*	29,2144
Sucrose*	51,6018

Source: (Adapted from Effiong et al. 2024)

Fatty Acids: the crude fat of edible mushrooms includes free fatty acids, mono-fatty acids, mono-fatty acids, mono-fatty acids, mono-fatty acids and mono-fatty acids. They also contain di- and triglycerides, sterols, sterol esters and phospholipids (Raman et al., 2021). In general, they have a low fat content. Some essential fatty acids are found in them, however, they are not known to be an important source of fatty acids to fulfil the needs of the human body. The main monounsaturated fatty acid is oleic acid (Table 04), and the main polyunsaturated fatty acid is linoleic acid, the latter being a precursor of the attractive flavour in dried mushrooms (Ewekeye et al., 2020) and present in greater quantities in this species. P. ostreatus contains n-6 essential fatty acids, oleic acid and linoleic acid in higher concentrations than other mushrooms. The amount of lipids can vary between 1.2% and 7.23% (Table 01).

Table 4. Fatty acid profile of Pleurotus ostreatus.

Fatty Acid	Value
Linoleic acid (C18:2)	80,59
Linolenic acid (C18:3)	0,2
Palmitoleic acid (C16:1)	0,09

3.2.2. Micronutrients

Oleic acid (C18:1) 9,1

Palmitic acid (C16:0) 9,18

Stearic acid (C18:0) 0,84

Source: (Adapted from Alsanad et al. 2021)

Vitamins: P. ostreatus is a rich source of various vitamins, especially B vitamins (Table 05). It contains significant amounts of thiamine (vitamin B1), riboflavin (vitamin B2), niacin (vitamin B3) and folate (vitamin B9). These vitamins are essential for energy metabolism, the formation of red blood cells and the maintenance of neurological function. In addition to the B vitamins, it also provides ergocalciferol (vitamin D2), which is crucial for bone health and the immune system (Gallotti; Lavelli, 2020).

Table 5. Vitamin profile of Pleurotus ostreatus.

Vitamin	Quantity
C16.46 mg/100g	
E21.50 mg/100g	
Beta carotene0.06 mg/100g	
A2,93 IU/100g	
B1 (Thiamine)0.6965 mg/kg	
B2 (Riboflavin)92.9696 mg/kg	
B3 (Niacin)0.7877 mg/kg	
B4 (Adenine).0047mg/kg B5 (Pantothenic Acid).6769mg/kg B6 (Pyridoxine) 0.6798 mg/kg	
B7 (Biotin)2.1117 mg/kg	
B8 (Inositol).2914mg/kg B9 (Folacin or Vit M)0.6897 mg/kg	
B10 (Para-aminobenzoic acid).6875mg/kg B11 (Folic acid)0.0534 mg/kg	
B12 (Cobalamin)0.3107 mg/kg	

Source: (Adapted from Effiong et al. 2024)

Minerals: the ash content reflects the presence of various minerals that play different roles in regulating the immune system, maintaining homeostasis, preventing disease and maintaining metabolic processes (Effiong et al., 2024). The high content of minerals such as Na, Ca, P, Fe and K (Table 06) in Pleurotus species makes it an alternative source to meat, fish and vegetables for replacing these elements (Raman et al., 2021). The amount of ash can vary between 3.74% and 11.21% (Table 01). However, elements such as iodine, fluorine, copper, zinc, mercury and manganese are also found in the fruiting bodies of mushrooms (Albuquerque et al, 2024).

Table 6. Mineral profile of Pleurotus ostreatus.

Mineral	Concentration (mg/kg)
Ca	0,08
Mg	7,00
K	12,25
In	1,93
Zn	2,73
Fe	9,66
Pb	0,33
Ni	0,23
Cd	0,08
Cr	0,02

Source: (Adapted from Effiong et al. 2024)

3.2.3. Bioactive compounds

The bioactive compounds found in mushrooms include polysaccharides, proteins, phenolic compounds, lignins, terpenes, beta-glucans, polyketides and some micronutrients. In certain concentrations, these compounds have antioxidant, immunomodulatory, anticancer, antiglycaemic, hepatoprotective, anti-hypercholesterolaemic, antibiotic and antifungal properties (Martínez-Flores et al., 2021). To promote each of these medicinal functions, many bioactive compounds act simultaneously (Table 07). For example, the main bioactive compounds with antioxidant capacity include vitamins (C and E), beta-carotene, ergosterol, lycopenes, phenolic compounds (diterpenes, flavanoids and volatile oils) and pleuran (polysaccharide). These bioactive compounds collectively contribute to the antioxidant properties of P. ostreatus, making it a functional food with potential health benefits. However, one substance can contribute to more than one medicinal function. Phenolic compounds have been linked to antioxidant activity, inhibition of lipid peroxidation, elimination of reactive oxygen species and iron ion chelating activity (González et al., 2021).

Table 7. Bioactive compounds of Pleurotus ostreatus and their medicinal functions.

Compounds	Medicinal Functions
Water-soluble protein or polysaccharide, ergothioneine	Anticancer
B-D glucan (pleuran), lectin, ergothioneine	Antioxidant
B-D glucan (pleuran), glycopeptides, proteoglycans, ergothioneine	Antitumour
Ubiquitin-like protein, ergothioneine	Antiviral
B-D glucan (pleuran)	Antibacterial
Lovastatin, ergotioneine	Anti-hypercholesterolaemic

Source: (Adapted from Lesa et al, 2022)

3.3. Cultivation

The global demand for mushrooms has grown significantly, driven by their nutritional appeal and the search for sustainable food, resulting in an increase in the number of mushrooms on the market. in production and consumption worldwide. Global production of mushrooms and truffles reached 48.34 million metric tonnes in 2022 (Shahbandeh, 2024), reflecting the economic and nutritional importance of these fungi. Among the most widely cultivated species are Lentinula edodes (shiitake), the most consumed mushroom in the world, followed by Pleurotus spp., Auricularia heimuer (black wood ear), Agaricus bisporus (mushroom), Flammulina filiformis (enoki mushroom) and P. eryngii (king oyster mushroom), which together produce more than one million tonnes per year (Li & Xu, 2022). Although Brazil is at a more recent stage of development in fungiculture, the country has significantly expanded its market share. A. bisporus, Pleurotus spp. and L. edodes mushrooms lead national production, cultivated by 52.2 per cent, 24.55 per cent and 16.44 per cent of producers, respectively (Marinho and Sousa, 2024). This growth reflects not only the increase in domestic consumption, but also Brazil's great potential to meet the growing global demand for mushrooms, which are valued for their functional properties and nutritional benefits.

The production of mushrooms by solid-state fermentation has proven to be an effective and sustainable strategy, especially for the reuse of agro-industrial by-products. In this method, the mycelium develops on sterilised organic substrates, free from other microorganisms, forming a network of cells that covers the surface of the solid medium (An et al., 2021). As well as replicating the fungus' natural environmental conditions, this technique promotes greater production efficiency at a low cost, as well as contributing to the circular economy by transforming agricultural waste into value-added products (Vaz Junior, 2020). In the Brazilian context, the production of Pleurotus spp. in particular has great growth potential, as the species can be grown on a variety of waste substrates, such as rice straw, sugar cane bagasse and waste from beer production and bean cultivation (Raman et al., 2021). These by-products, which are often discarded, offer a sustainable alternative for mushroom production, contributing to both the reduction of pollution and the conservation of natural resources. The use of this waste in fungiculture not only helps to mitigate environmental impact, but also adds value to little-explored production chains, driving the development of new business models in Brazil and around the world.The cultivation of P. ostreatus is divided into different stages:Preparation and handling of spawn: spawn is a substrate completely colonised by the fungus, used to propagate the mycelium in the desired solid substrate during cultivation. There are four main categories o f spawn for mushroom cultivation: sawdust spawn, grain spawn, liquid spawn and stick spawn (Zhang et al., 2019). The most common support matrix is made up of sterilised grains, such as soya, wheat, maize, sorghum and rice, due to their biochemical properties and practical performance. Mushrooms are propagated by means of mycelium, which grows on a sterilised solid substrate to avoid contamination by unwanted microorganisms and guarantee the genetic purity of the fungus. When the mycelium is added to the substrate to produce the "spawn", it quickly colonises the substrate, which can then be stored and transported to inoculate other substrates, transforming them into favourable media for mushroom development. The success of mushroom production depends largely on the quality of the spawn. A healthy and vigorous spawn ensures rapid colonisation of the substrate and efficient mushroom production (Sánchez, 2010).

Substrate preparation: substrates cannot be inoculated directly with fungi, but require prior pasteurisation/sterilisation. Heat treatment of the substrate is a crucial step in mushroom cultivation, as it creates the ideal conditions for the colonisation and development of the fungi, while inhibiting or reducing the growth of pathogens and competitors that could interfere with cultivation. For example, Yang et al. (2022) suggested sterilisation under high pressure (121-123 °C for 2-3 hours) or sterilisation under atmospheric pressure (95-100 °C for 8-12 hours) as pre-treatments for mushroom growing media. However, a more sustainable procedure for achieving pasteurisation of substrates consists of their prior short-term composting (around 7-10 days), taking advantage of the natural thermophilic phase of this process (De Mastro et al., 2023).Inoculation: after the substrate has been properly prepared, it is inoculated with the "mushroom spawn" produced earlier. An inoculation rate of 10 per cent is recommended as adequate to achieve the best results (Attaran Dowom et al., 2019). Inoculation must be carried out in an environment free from contamination, and the equipment used must be properly sterilised.Incubation: after inoculation, the substrates are incubated preferably in the dark (Zheng et al., 2021), at a temperature ranging from 25° to 30°C (Albuquerque et al., 2024) and ambient humidity ranging from 80% to 95% (Wang et al., 2020). During this period, the mycelium develops in the substrate, secreting enzymes that degrade the organic matter present, transforming it into nutrients that can be absorbed by the fungi.Formation of primordia: in response to specific environmental changes, such as variations in temperature, CO levels$_2$, O availability$_2$ and light intensity (Sydor et al., 2022), the mycelium begins to form small structures called primordia, which eventually develop into basidiomata, which are the mushrooms themselves.Harvesting: mushroom cultivation time depends on the formulation used for cultivation (Tavarwisa et al., 2021). Once they have reached the desired size and maturity, usually with the hats open and the edges rolled down, the mushrooms are harvested by hand, twisting them at the base or cutting them with a sharp knife. This process can be repeated if the substrate used for cultivation provides the necessary nutrients.Post-harvest: after harvesting, mushrooms can be processed for immediate consumption in natura, dehydrated or preserved for future use. When properly preserved, the shelf life can reach up to 15 days without considerable losses in flavour and nutritional composition (Rahman et al., 2020).

3.4. Co-products used in the production of Pleurotus ostreatus

The substrates used in the production of P. ostreatus can be classified into three large groups: food processing by-products, agro-cereal by-products and fruit and nut by-products (Doroski et al., 2022).Some examples of substrates used to grow P. ostreatus include rice straw, cereal straw, wheat straw, barley straw, maize straw, sugar cane bagasse, maize stalk coproduct, wheat stalk, cotton coproduct and rice husks, banana leaves, elephant grass, bamboo leaves, soya straw, Triplochiton scleroxylon sawdust, beech sawdust and flax tow (Figure 08) (Raman et al., 2021). The nutritional composition of each of these materials is different, however, they all have the necessary combination of nutrients to provide mycelial growth and the formation of fruiting bodies. In general, the quantities of nutrients needed to grow P. ostreatus vary, as shown in Table 08.

Figure 8. Main co-products used in the cultivation of Pleurotus ostreatus var. florida.

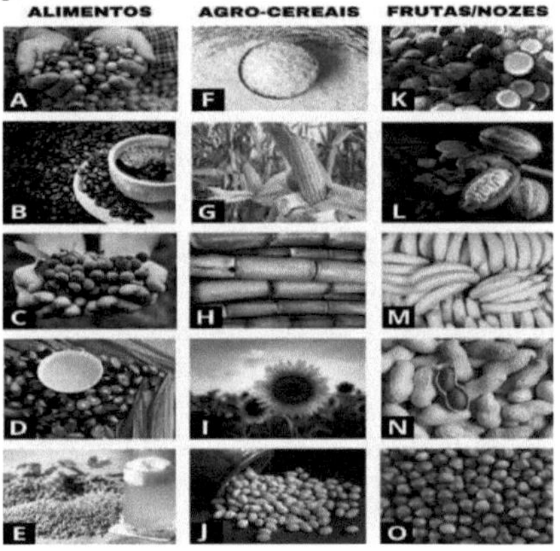

A = Olive, B = Coffee, C = Grape, D = Palm, E = Beer, F = Rice, G = Maize, H = Sugarcane, I = Sunflower, J = Soya, K = Coconut, L = Cocoa, M = Banana, N = Peanut, O = Hazelnut.
Source: Own authorship, 2024.

Table 8. Suitable composition of formulations for growing Pleurotus ostreatus.

Property	Nutritional Composition
Lignin (%)	~11-15
Cellulose (%)	~29-35
Hemicellulose (%)	~7
Protein (mg/g)	~18-24
Polysaccharide (mg/g)	~7-10
Carbon (%)	~32-37
Hydrogen (%)	~4-5
Nitrogen (%)	~1-3
Oxygen (%)	32-34
Humidity (%)	~42-56
Cu (mg/kg)	13.4
Zn (mg/kg)	14.2
Pb (mg/kg)	7.7
Cd (mg/kg)	0.41
Ni (mg/kg)	5.5
	Adapted from (Wan Mahari et al., 2020)

3.4.1. Co-products brewers

The brewing process is divided into different stages: malting, milling, mashing, boiling, fermentation, maturation, bottling and pasteurisation (Vivian et al., 2016). In malting, the barley grain is germinated to stimulate enzyme synthesis and starch degradation. This is followed by milling, where the malt is crushed to release the starch. During mashing, the grain is mixed with water and heated to convert the starch into fermentable sugars. At the end of this process, the liquid is filtered to obtain the must, a sugary solution. In the boiling stage, the wort is boiled with hops to extract bitterness and aroma. After separation and cooling, the liquid is fermented. After a few days' rest, the fermented liquid is stored to promote carbonation and improve flavour during maturation. Finally, the beer is bottled and pasteurised to complete the process (Figure 09).

Figure 9. Brewing process.

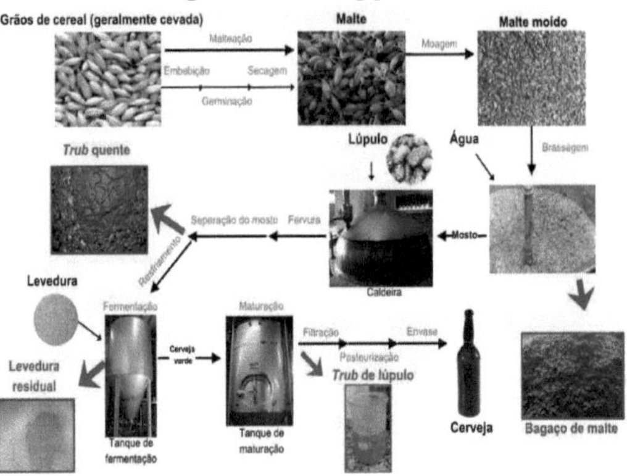

(Adapted from Rachwal et al., 2020).

The brewing process generates a large quantity of co-products with high nutritional potential (Aguiar et al., 2022; Rodriguez et al., 2023). The production of 1,000 tonnes of beer results in approximately 10,000 tonnes of liquid and 137 to 173 tonnes of solid by-products (Amoriello; Ciccoritti, 2021). The main co-products generated are malt pomace, hot and cold trub and residual yeast.Malt pomace, removed after the mashing stage, accounts for 85% of total co-products, with around 14 to 20 kg discarded for every hectolitre of beer produced (Amoriello et al., 2020). Trub, which can be classified as hot (or thick) and cold (or thin), is removed after boiling and during fermentation/maturation respectively, accounting for approximately 0.2 to 0.4 kg per hectolitre of beer produced (Olajire, 2020). Hop trub is the result of the purging of solid material from the bottom of the fermenter after the cold hopping or dry hopping stage, which consists mostly of hops, as well as containing beer and probably yeast in smaller quantities. The residual yeast, collected after fermentation, corresponds to around 0.3 kg per hectolitre of beer produced (Karlović et al., 2020).The composition of these co-products can vary according to the type of beer produced. However, some major components can be highlighted. Pomace Malt (Figure 10) is a lignocellulosic material, composed mainly of fibres (hemicellulose and cellulose), protein and lignin. Fibres make up around half of the composition of malt cake by dry weight, while proteins can make up to 30%. The chemical composition of malt cake includes hemicellulose (20-25%), cellulose (12-25%), lignin (12-28%), protein (19-30%), lipids (10%) and ash (2-5%) (Lynch et al., 2016). This composition, rich in hemicellulose, cellulose and lignin, makes malt pomace a nutritionally complete co-product that can be used as a base substrate for growing commercial fungi, particularly for the production of edible mushrooms.

Figure 10. Malt bagasse

Source: https://l1nk.dev/wHzL0

The hot trub (Figure 11) is mainly made up of water, accounting for around 80 per cent of its mass. In addition, around 85% of the hop particles used in brewing do not dissolve in the wort and are discarded along with the hot trub. The unstable, high molecular weight colloidal proteins coagulated during the boiling of the wort (Costa et al., 2021). These proteins, originating from both the malt and the adjuncts used in the process, correspond to a significant fraction, ranging from 40% to 70% of the total dry weight of the hot trub (Saraiva et al., 2019). It also contains complex carbohydrates, lipids, minerals, tannins, small malt particles and a C/N ratio of 6.3, which is considered low (Rachwal et al., 2020). As a co-product composed mainly of water and proteins, as well as having phenolic compounds with antioxidant capacity, hot trub has the potential to be used as a supplement in the cultivation of edible mushrooms.

Figure 11. Filtered hot trub

Source: https://acesse.one/tWiS0

Residual yeast (Figure 12) has a high protein content (45-60% of dry weight), with approximately 64% of this total made up of essential amino acids (Jaeger et al., 2020). It also has a high moisture content (74-86%), a high carbon content (45-47% of dry matter), a C/N ratio between 5.1 and 5.8, and is rich in phenolic compounds and B vitamins (B1, B2, B3, B6 and B8) (Rachwal et al., 2020). Its mineral content includes sodium, potassium, calcium, magnesium, iron, zinc, manganese and copper, as well as phenolic acids such as gallic acid, catechin, protocatechuic acid, p-coumaric acid, ferulic acid and cinnamic acid (Jaeger et al., 2020).

Figure 12. Residual yeast

Source: https://l1nk.dev/joSeR

Hop trub (Figure 13) is a brewing co-product that has not yet been explored and studied. This co-product, generated during the maturation stage, can contain proteins, lipids, organic acids and fine particles of hops and malt left over from the brewing processes. previous. The available literature is scarce on the physicochemical characteristics of hop trub, which limits the understanding of its potential as a supplement in a context such as mushroom cultivation.

Figure 13. Hop trub

Source: Authors, 2024

3.4.2. Bean pods from

Beans (Phaseolus vulgaris L.) are one of the main components of the Brazilian population's diet. Brazil is the world's largest producer and consumer of this food (Barbosa et al., 2021), especially in the Northeast, which produces approximately 788,000 tonnes of green beans per year (Coelho; Ximenes, 2020). According to data from Conab's twelfth survey of the Brazilian grain harvest, bean production in the 2021/22 harvest was 2.997 million tonnes (Marinho and Sousa, 2024), most of which is destined for domestic trade. For example, in 2017, around 97% of Brazil's bean supply was destined for the domestic market (Barbosa et al., 2021). Bean pods (Figure 14) are a source of organic acids, phenolic compounds, vitamins, minerals, proteins, fibres and sugars (Omura et al., 2020). It has an average moisture content of 87.31% and, by dry mass, around 33% cellulose, 46% hemicellulose, 17% lignin and 29% protein (Kalili et al., 2022). This nutritional composition makes the green bean pod an excellent nutritional source and a suitable co-product for application as a base substrate in the cultivation of fungi of commercial interest, with potential for the production of edible mushrooms (Albuquerque et al, 2024).

Figure 14. Bean pods and grains (Phaseolus Vulgaris L.).

Source: http://agroresidues.hg.

3.5. Factors influencing cultivation

The growth and differentiation of fungi are complex processes that depend on multiple variables, which exert different simultaneous influences. The factors involved in mushroom production include extrinsic and intrinsic factors. Extrinsic factors include light, temperature, air composition, environmental humidity, the amount of spawn, and characteristics of the cultivation bioreactor, such as the opening area and the arrangement of the holes. Intrinsic factors include pH, particle size, substrate humidity and the composition of the growing medium. These factors not only affect the dimensions of the mushrooms, technically referred to as biological parameters, but also influence the final production parameters and nutritional composition of the mushrooms (Figure 15). Although they represent different concepts and measures, in practice, their values are interrelated, as biological parameters not only influence production parameters, but also affect nutritional composition (Zakil et al., 2019; Wan Mahari et al., 2020).

Figure 15. Factors affecting nutritional composition, biological parameters and production parameters in mushroom cultivation.

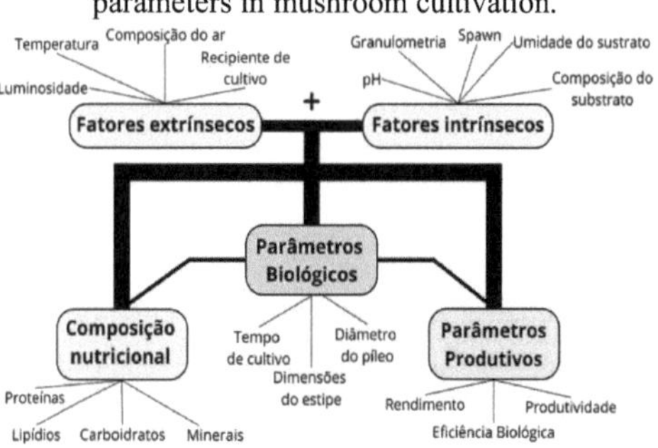

Source: authorship, 2024.

3.5.1. Extrinsic factors

3.5.1.1. Luminosity

Light plays a critical role in the development of fungi, influencing their morphological, reproductive and productive processes. Fungi are sensitive to different wavelengths due to specific photoreceptors, such as flavin receptors for blue light, retinal receptors for green light and tetrapyrrole receptors for red light (Bayram; Bayram, 2023). Morphological and reproductive changes are triggered by the presence or absence of light (Li et al., 2023). Studies have shown that different colours of light significantly impact the growth of Pleurotus spp. mycelium. Red light promotes higher radial growth compared to the control (Kaur et al., 2022). Under green light conditions, an increase in the enzymatic activity of cellulase, endoglucanase, xylanase and lacase was observed in relation to the mycelial biomass of P. ostreatus (Araújo et al., 2021), indicating a direct influence of wavelength on the growth phase. In addition to mycelial growth, light also affects the formation of fruiting bodies. Blue light favours the development of fruiting bodies in P. ostreatus, while red light can have varying effects, including negative effects on the induction of these structures (Wang et al., 2020). Light intensity also plays a role This is a crucial factor in productivity, with higher levels being associated with an increase in the total yield of fruiting bodies (Yue et al., 2022). Therefore, single-frequency light or mixed treatment comprising red and blue light in different proportions can be used to improve mycelial growth and produce more petioles and stipes.In terms of reproduction, the duration and direction of exposure to light influence sporulation and mycelial growth. The continuous presence of light favours sporulation, while dark environments are ideal for mycelial growth and substrate degradation, because the presence of light conveys the signal of an open area, suitable for fungal dispersal, while the lack of light indicates an airtight space (Zheng et al., 2021). In addition, the direction and intensity of the light can affect the orientation of the stipe and pileus, demonstrating phototropism (Albuquerque; Sousa, 2024).

3.5.1.2. Temperature

Temperature has a significant influence on the mycelial growth rate of fungi, directly impacting colony morphology, reproduction (Zheng et al., 2021) and effectiveness in degrading the substrate through the action of enzymes essential for the development of the mycelium, such as lacases, xylanases, amylases and proteases (Albuquerque; Sousa, 2024).
In addition, temperature affects the dimensions of the mushrooms, impacting the final yield. Lower temperatures tend to result in smaller pileus diameters and stipe heights (Enoch Peter, 2019), which reduces the total yield. However, excessively high temperatures compromise biological efficiency, requiring a thermal balance to optimise results. Pleurotus spp. are well adapted to temperatures in tropical and subtropical climates, ranging from 25°C to 30°C (Albuquerque et al, 2024). During the vegetative stage, the activities of substrate-degrading enzymes such as lacase, xylanase and cellulase are higher at lower temperatures, between 15°C and 22°C (Hu et al., 2023). However, the optimum temperature for mycelial growth and fruiting body formation is 28°C (Enoch Peter, 2019), providing a faster mycelial growth rate and more efficient primordia formation.

3.5.1.3. Composition of air

Carbon dioxide (CO_2) plays a crucial role in the development of mushroom fruiting bodies. Its concentration in the environment can significantly affect various stages of fungal growth and development. In the early phase of development, mushrooms are particularly sensitive to CO_2 , which can negatively affect cell morphology and the synthesis of cell wall components such as R-glucan (Lin et al., 2022). At high concentrations, CO_2 inhibits the differentiation of the pileus (mushroom cap) and can result in malformed fruiting bodies with long, thin stipes (Bellettini et al., 2019). Concentrations of CO_2 higher than 0.1% can be toxic to fruiting bodies. When the concentration of CO_2 reaches 5%, pileus differentiation is inhibited and the formation of fruiting bodies is severely affected (Lin et al., 2022). During growth in the dark, it is important to keep the CO_2 concentration between 2000 and 2500 mg/L. For the formation of fruiting bodies, the ideal concentration is between 1500 and 2000 mg/L (Bellettini et al., 2019).Aerobic mushrooms require oxygen for their survival and development. The concentration of oxygen influences both mycelial growth and the production of fruiting bodies. Under suitable oxygen conditions, Pleurotus spp. mycelium grows vigorously, colonising the substrate efficiently. A lack of oxygen inhibits mycelial growth, resulting in less dense colonies and slower development. During the formation of primordia, the respiratory activity of mushrooms increases, requiring large amounts of oxygen to metabolise CO_2 during growth (Lin et al., 2022). Therefore, in the fruiting phase, a reduction in the concentration of CO_2 , as well as an increase in O_2 (Bellettini et al., 2019).On the other hand, the relative humidity of the environment is an essential factor in mushroom cultivation. Adequate humidity is crucial to avoid problems such as mushroom dehydration, undesirable microbial growth and discolouration. Proper humidity management ensures a favourable environment for the healthy development of mushrooms, promoting more efficient and better quality production. Excess humidity, for example, favours the condensation of water vapour on the surface of the mushrooms, which accelerates undesirable microbial growth and discolouration, thus affecting mushroom growth and productivity (Wan Mahari et al., 2020). One of the most effective methods of maintaining optimal humidity for production is irrigation around the cultivation bioreactors. Pleurotus spp. show greater mycelial growth at humidities between 60% and 80% (Bellettini et al., 2019), while humidity between 80% and 95% favours the formation of the fruiting body (Wan Mahari et al., 2020). Saputera et al. (2020) observed that the best irrigation frequency for greater growth and yield of oyster mushrooms is four times a day: morning, afternoon, evening and dawn.

3.5.1.4. Bioreactor

The bioreactors, characterised as polyethylene plastic bags, influence both the size of the mushrooms and the final yield. Perforated bioreactors produce a superior and earlier harvest compared to non-perforated bioreactors, due to the accumulation of CO2 which inhibits fruiting. Iryna et al. (2021) observed a linear correlation between the size of the perforations and the size of the clusters of mushroom fruiting bodies. In that study, 150 mm perforations resulted in a higher final weight of the mushrooms. In addition, the total yield of fruiting bodies and the biological efficiency (BE) in bioreactors placed in the vertical and inclined position were 10 per cent higher than in the horizontal position. The number of holes in the bioreactors is also an important factor to consider during cultivation. An increase in the number of holes in the bioreactors results in smaller mushrooms (Bellettini et al., 2019). Therefore, the proper choice of the size and number of holes, as well as the storage position of the bioreactors, are essential to optimise mushroom growth and productivity.

3.5.2. Intrinsic factors

3.5.2.1. pH

pH can influence cell membrane function, morphology, cell structure, nutrient absorption and biomass production (Attaran et al., 2019). In addition, pH can affect the solubility of nutrients in the substrate, especially minerals essential for proper mushroom growth, and can also influence the activity of enzymes involved in fungal metabolism, which are crucial for the degradation of the raw material and the assimilation of nutrients (Albuquerque et al, 2024).
The pH of substrates must be in the range of 3 to 10 for mushrooms to grow in tropical regions (Zakil et al., 2019). For Pleurotus spp., the ideal pH can range from 4 to 6 (Vilas et al., 2020). However, each stage of production has different pH requirements. For example, a pH between 4.0 and 7.0 is ideal for the growth of the mycelium, while a pH between 3.5 and 5.0 is more suitable for the formation of the fruiting body (Bellettini et al., 2019).

3.5.2.2. Humidity of the substrate

The humidity of the substrate is essential for the growth of the network of filaments that make up the structure of fungi and directly affects their expansion and development. Water aids in the exchange of gases and nutrients, as well as being indispensable for the fungus's metabolism, since the cell wall is degraded through hydrolysis and the production of hydrogen peroxide is important in the degradation of lignin bonds (Aghajani et al., 2018). Substrates with a higher moisture retention capacity perform better than those with a lower moisture retention capacity (Fufa et al., 2021). It has been shown that P. ostreatus changed degradation selectivity when the moisture content of the wood decreased from 65% to 45%, showing that moisture also affects enzyme expression (Aghajani et al., 2018). Lack of moisture can lead to the formation of shrivelled mushrooms, with irregular growth or deformations, and even death. In contrast, excess moisture reduces the porosity of the substrate, resulting in limited oxygen transfer, difficulty in respiration for the mycelium, inhibiting transpiration and making it impossible for fruiting bodies to develop (Bellettini et al., 2019). In addition, high levels of humidity can increase the risk of problems such as rotting and bacterial or fungal contamination (Navarro et al., 2020). The ideal humidity for growing Pleurotus spp. is between 65% and 75%. Maintaining this humidity range is crucial to ensuring healthy growth and high mushroom productivity.

3.5.2.3. Granulometry

The efficiency of enzymatic degradation is influenced by the contact surface of the substrate, which directly affects the development of the mycelium. Smaller particles, i.e. a larger contact surface, favour the degradation of the substrate by the action of the enzymes (Soh et al.,

2021). However, if the particles are too small, the wet substrate can become excessively compacted, reducing the available porosity and aeration (Zakil et al., 2019). Maintaining the right balance in substrate granulometry is crucial to ensuring efficient degradation and a favourable environment for mycelium growth. Substrates with optimised granulometry promote better mycelium penetration and facilitate gas exchange, resulting in more vigorous growth and greater mushroom productivity.

3.5.2.4. Spawn, seed or inoculum

The inoculation rate of the substrate is a crucial factor that accelerates mycelial growth and consequently influences the total production time. A suitable inoculation rate can optimise mycelium development, improve biological efficiency and reduce the risk of contamination. From an economic and practical point of view, an inoculation rate of 10 per cent is recommended as ideal. Inoculation rates of less than 10% can facilitate contamination of the substrate and reduce biological efficiency, jeopardising crop productivity (Attaran Dowom et al., 2019). In addition, insufficient inoculation can prolong the colonisation time of the substrate, resulting in a longer and less efficient production cycle.

3.5.2.5. Composition of substrate

The composition of the substrate plays a crucial role in the development of fungi, influencing their growth both positively and negatively. High levels of lipids can alter the morphology of the mycelium, increasing hydrophobicity, which is not favourable for healthy growth (Soah et al., 2021). Substrates rich in inorganic substances and cellulose are considered ideal for growing Pleurotus spp. (Zakil et al., 2022). However, the lignin content can negatively impact the yield of fresh mushrooms due to its more difficult degradation by the enzymes produced by Pleurotus spp. (Soh et al., 2021).The carbon and nitrogen content of the substrate is critical for mycelial development and fruiting. Wan Mahari et al. (2020) recommend a nitrogen content of between 1.84% and 2.08% and a carbon content of around 40% for best production results. Carbon sources such as glucose, dextrose, sucrose and molasses are metabolised as cellular energy to support mushroom growth, while nitrogen sources include nitrates, organic nitrogen compounds and ammonium ions. A high C/N ratio favours mycelial growth, while a low ratio favours basidiocarp formation (Hoa et al., 2015). Supplementation with external nitrogen sources is recommended to increase the production of P. ostreatus (Raman et al., 2021). Studies show that C/N ratios between 7:1 and 40:1 are associated with higher yields (Albuquerque and Sousa, 2024). The presence of minerals in the substrate also has a significant influence on productivity. For Pleurotus ostreatus to grow properly, it is essential that the substrate contains minerals such as sulphur (S), calcium (Ca), magnesium (Mg), phosphorus (P), potassium (K) and other elements such as iron (Fe), zinc (Zn), manganese (Mn), copper (Cu) and molybdenum (Mo) (Cueva et al., 2017). The adequate concentration of these elements is directly related to the efficient degradation of cellulose, hemicellulose and lignin. The presence of copper (Cu), beryllium (Be), magnesium (Mg), manganese (Mn), iron (Fe), lithium (Li) and strontium (Sr) is positively correlated with the degradation of hemicellulose, increasing biological efficiency. On the other hand, elements such as cadmium (Cd) and strontium (Sr) have adverse effects on cellulose degradation, and cadmium (Cd), chromium (Cr), nickel (Ni) and lead (Pb) can reduce the expression of ligninolytic enzymes, impairing productivity (Koutrotsios et al., 2020). In addition, calcium (Ca) and magnesium (Mg) in the culture medium are associated with a higher dry weight of the mycelium, due to their roles as cofactors in essential metabolic processes such as glycolysis, respiration and protein synthesis (Zhou; Parawira, 2022).

All these factors (extrinsic and intrinsic) also influence the nutritional composition of mushrooms. UV light can significantly increase the Vitamin D2 content in P. ostreatus (Gallotti; Lavelli, 2020). The study by Zawadzka et al. (2022) showed that P. ostreatus grown under white light of 200 lux had higher riboflavin (vitamin B2) levels compared to mushrooms exposed to lower intensity light. When adjusting pH, the addition of hydrochloric acid was associated with a reduction in carbohydrate content, while increasing the amount of protein in the fruiting bodies (Nwoko et al., 2021). The composition of the substrate in relation to a higher amount of nitrogen is correlated with the protein content in the basidiomes (Raman et al., 2021). A lower C/N ratio (carbon/nitrogen) favours high levels of crude protein, amino acids and nucleotides, as well as influencing the production of bioactive compounds (Hoa et al., 2015; Atlh et al., 2019). In addition, the concentration of ash in the substrate also influences the final mineral and protein content of the mushrooms (Koutrotsios et al., 2020).

4. MATERIAL AND METHODS

4.1. Origin of Mushroom

The matrix of P. ostreatus var. florida belongs to the fungal culture collection of the Molecular Genetics and Plant Biotechnology Laboratory and is being maintained on 2 % Sabouraud-dextrose agar medium (5 g/L meat peptone, 20 g/L glucose, 5 g/L casein peptone and 15 g/L bacteriological agar), according to the methodology described by Furlan et al. (1997).

4.2. Types and origins of substrates and supplements used

Malt bagasse (Hordeum vulgare L.) (Vierbrauer Brewery, Cabedelo/PB) and green bean pods without the grain (Phaseolus vulgaris) (Torre Public Market, João Pessoa/PB) were used as substrates. Hot trub, hop trub and residual yeast (SafLager™ W-34/70) (Cervejaria Vierbrauer, Cabedelo/PB) were used as supplements. Popcorn maize (Zea mays var. everta) was used for seed production (Commercial origin).

4.3. Processing of Substrates

4.3.1. Pomace from malt

It did not receive any specific treatment, and due to its humidity, no distilled water was added during preparation. After collection, the pomace was packed in plastic bags and kept refrigerated at 4°C.

4.3.2. Bean pods without the beans

It was first dehydrated at 70°C for 24 hours in a drying oven. Next, it was submerged in tap water for 24 hours and, after draining the water, it was placed on a flat platform for 2 hours to dry. Finally, the pods were ground in a knife mill and packed in plastic bags kept refrigerated at 4°C.

4.3.3. Residual yeast, hot trubs and hops

They did not receive any specific treatment. After collection, they were placed in plastic buckets and kept refrigerated at 4°C.

4.3.4. Corn from popcorn

The beans were cooked over high heat in a pressure cooker for 90 minutes. After cooking, the extract from the cooking process was drained off and the grains were left at room temperature (28 ± 2°C) to dry externally for approximately 60 minutes. They were then packed in plastic bags and kept refrigerated at 4°C (Bonatti et al., 2004).

4.4. Production of seed:

Initially, the processed popcorn was placed in a glass jar (6.0 x 8.5 cm) filling 80% of its volume. The jar was then closed with aluminium foil and autoclaved at 121°C for 30 minutes. After cooling to room temperature (28° ± 2°C), three discs of mycelium with a diameter of 5 ± 1 mm were inoculated into the flask. Incubation took place at 28° ± 2°C, in the absence of light, until the surface was completely colonised.

4.5. Formulations for the production of Pleurotus ostreatus var. florida

100g of substrates T1 (Bean pods without grains, 80%) + Malt bagasse, 20%), T2 (T1 + residual yeast, 5%), T4 (T1 + Hot Trub, 5%) and T3 (T1 + residual yeast, 10%).
%) , T4 (T1 + Hop Trub, 10 %) were weighed and, where necessary, mixed with their respective supplements (Table 09). The treatments were placed in the bioreactors, characterised as polyethylene plastic bags (12 x 20 cm), partially closed with polyamide ties and autoclaved at 121°C for 60 minutes (Figure 14). After 24 hours, 3 corn kernels with adhered mycelium were added to each bioreactor. The treatments were incubated at 28° ± 2°C in the absence of light for 15 days. An experiment was carried out with 10 replicates. After incubation, the bioreactors were opened to assess the fruiting potential. The fruiting bodies were harvested when they showed maximum development, as seen by the beginning of the unwinding of the pileus margins.

Table 09: Formulations for growing Pleurotus ostreatus var. florida.

Table 9. Formulations for growing Pleurotus ostreatus var. florida.

Treatments	Formulations and proportions
T1	Bean pods without beans - VF (80 %) + Malt bagasse - BM (20 %)
T2	Bean pods without beans - VF (80 %) + Malt bagasse - BM (20 %) + residual yeast - LV* (5 %)
T3	Bean pods without beans - VF (80 %) + Malt bagasse - BM (20 %) + hot trub - TBQ* (10 %)
T4	Bean pods without beans - VF (80 %) + Malt bagasse - BM (20 %) + hop trub - TBL*(10 %)

***The** percentage added is calculated on 100 per cent of the substrate mass. **VF:** Bean pods; **BM:** Malt bagasse; **LV:** Yeast**; TBQ:** Hot trub; **TBF:** Hop trub. Source: Authors, 2024.

4.6. Evaluation of parameters biological

During the cultivation period, biological aspects of the mushroom were analysed, such as: time taken to colonise the substrate, period of primordia emission, time taken to form basidiocarps, height and width of the stipe (cm) and diameter of the pileus (cm).

4.7. Evaluation of production parameters

After harvesting the mushrooms, production parameters were analysed according to the growing substrate, such as: mushroom and treatment moisture (U%), yield (R%), biological efficiency (EB%), productivity (Pr g/day^{-1}), loss of organic matter (PMO %) and compost consumption (CC%).

4.7.1. Income

It was measured in terms of the fresh weight (g) of the mushrooms produced per bioreactor containing 100g of compost (Zakil et al., 2022).
Where: MCU = Wet mass of fruiting bodies, g; MSSI = Dry mass of initial substrate, g.

$$R(\%) = \frac{MCU}{MSSI} \times 100 \quad \text{Eq. 01}$$

4.7.2. Biological Efficiency (EB%)

The biological efficiency (EB%) of the process was determined by the ratio between the mass of fresh fruiting bodies and the mass of substrate based on their dry weight according to Equation 02 (Muswati et al., 2021).

$$EB(\%) = \frac{MBS}{MSS} \times 100 \quad \text{Eq. 02}$$

MBS = Mass of fresh basidiocarps, g MSS = Dry mass of substrates, g

4.7.3. Productivity

The productivity (Pr g/day^{-1}) of the process was determined according to HOLTZ (2009). It consists of the ratio between the mass of dried fruiting bodies and the total cultivation time (time from inoculation to production flow, as shown in Equation 03.

$$\text{Pr}(g/dia) = \frac{\text{MCS}}{\text{TC}} \quad \text{Eq. 03}$$

Where: MCS = Dry mass of fruiting bodies, g; TC = Total cultivation time, days.

4.7.4. Loss of organic matter

Index that evaluates the decomposition of the substrate by the fungus. This index is based on the loss of organic matter decomposed by the fungus, which is determined by the difference between the dry mass of the initial substrate and the dry mass of the residual substrate (Equation 04) (RAJARATHNAM; BANO, 1989).

$$PMO\ (\%) = \frac{(MSSI - MSSR) \times 100}{MSSI}$$
Eq. 04

In which:

PMO= Loss of organic matter, %
MSSR= Residual substrate dry mass, g MSSI= Initial substrate dry mass, g

4.7.5. Consumption percentages of compost

The index was calculated based on the dry mass of the substrate after cultivation divided by the initial dry mass of the compost, as shown in Equation 05.

$$CC(\%) = \frac{CF \times 100}{CI}$$
Eq. 05

Where CF = Final dry matter compost; CI = Initial dry matter compost.

4.7.6. Humidity of treatments and mushrooms

Moisture was determined by the gravimetric method using heat, which is based on the weight loss of the material when heated until it reaches a constant weight. All the treatments were kept in a drying oven at 105 °C for 24 hours and the moisture content of the substrate was determined by the difference between the wet and dry biomass (Equation 06).

$$U(\%) = \frac{(MI - MF) \times 100}{MI}$$
Eq. 06

Where: MI = Initial mass of the sample; MF = Final mass of the sample.

4.8. Physico-chemical analysis of raw materials

Malt bagasse and bean pods without grains were analysed individually using 3 replicates of each substrate for the analyses of reducing sugars, non-reducing sugars, total sugars, minerals, pH and total soluble solids (ºBrix). For the analyses, all the substrates were crushed in a knife mill, dried in a LUCA-80/27 oven at 102º C and dried in a drying room. °C for 24 hours and stored at room temperature between 25°C and 32°C. In addition, the carbon and nitrogen content of malt cake and bean pods without the grain were analysed.

4.8.1. pH

The pH of the substrates was determined using a bench pH meter, previously calibrated with buffer solutions of pH 7 and pH 4. For the measurement, 5 g of each dry substrate was suspended in 50 mL of distilled water. The suspension was homogenised using a magnetic stirrer before the pH reading was taken.

4.8.2. Ashes

Minerals were determined by carbonising 5g of the material in triplicate in a muffle furnace at 550°C to decompose all the organic parts and the results were expressed in % according to the methodology of Nogueira and Souza (2005).

4.8.3. Sugars reducing

To analyse the substrates, 10g of the sample was diluted and filtered with 250 mL of distilled water, 25 mL of the filtrate was transferred to a burette and 10 mL of Fehling's solutions A (copper sulphate) and B (sodium potassium tartrate and sodium hydroxide) were pipetted into a 250 mL Erlenmeyer flask, topped up with distilled water and heated to boiling point, which was kept constant during the titration until a brick red colour appeared. Methylene blue 1% was used as an indicator. Sugars were measured as a percentage using Equation 06:

$$\%Acúcares\ redutores = \frac{F \times C \times 100}{P \times V} \quad \text{Eq. 06}$$

F = Fehling's liquor factor (F/2 when using 5 mL)
C = volumetric capacity of the flask used (250 mL) P = sample weight
V = volume spent on titration
Solutions with a pH below 6 were neutralised with 0.1N NaOH. And samples with intense colours were clarified with 1.0M zinc acetate and 10 mL of 0.25M potassium ferrocyanide. To determine the reducing sugars in the supplements, 0.5 mL of the sample and 0.5 mL of DNS (3,5-dinitrosalicylic acid) solution were used. The mixture was The solution was heated in a water bath at 100 °C for 5 minutes. Then 4 mL of distilled water was added and the absorbance of the resulting solution was measured at 540 nm. The procedure was repeated for all the samples, and the results were plotted on a graph of absorbance (y-axis) versus concentration (x-axis), using the straight line equation to determine the concentration of reducing sugars.

4.8.4. Sugars total

The total sugars of the substrates were determined using 5 g of the sample diluted and filtered with 250 mL of distilled water, 10 mL of concentrated HCL were added and the solution placed in a water bath at 68 to 78 °C for 20 minutes. After cooling, the solution obtained was transferred to a 25 ml burette, 5 ml of Fehling's solutions A and B were volumetrically pipetted into the solution, distilled water was added and heated to boiling. The sample was titrated while boiling until a brick-red colour appeared, using 1% methylene blue as an indicator in accordance with the Lane-Eynon method (2008). Total sugars were calculated using Equation 07:

$$\%\ Açúcares\ totais = \frac{(F \times C \times 100)}{(P \times V)} \quad \text{Eq. 07}$$

In which:
F = Fehling liquor factor
C = volumetric capacity of the flask used - 250 mL V = volume spent on titration
To analyse the total reducing sugars in the supplements, 0.5 mL of the sample and 0.5 mL of 2 M HCl were added to a test tube. The mixture was heated in a water bath at 100 °C for 5 minutes to hydrolyse the non-reducing sugars. After this time, 1.5 mL of 1 M NaOH was added to neutralise the solution. Then 0.5 mL of this neutralised solution was transferred to another test tube and mixed with 0.5 mL of DNS solution. The mixture was heated in a water bath at 100 °C for 5 minutes and, after cooling, 4 mL of distilled water was added. The absorbance was measured at 540 nm to quantify the total reducing sugars.

4.8.5. Non reducing sugars

Non-reducing sugars are classified due to the absence of free aldehyde and ketone groups. They are not capable of reducing salts and need to undergo hydrolysis of the glycosidic bond in order to oxidise. Non-reducing sugars were calculated using the difference between total sugars and reducing sugars according to Equation 08.

$$\%\text{Açúcares não redutores} = \text{Açúcares totais} - \text{Açúcares redutores}$$

Eq. 08

4.8.6. Total soluble solids (°Brix)

A 2 g sample of each dry substrate was diluted with 15 mL of distilled water and homogenised. The large particles were then discarded. Afterwards, 1 to 2 drops were transferred to the prism of the RHB0-90 analogue refractometer for % Brix and read on the scales in °Brix.

4.9. Physico-chemical analysis of mushrooms

Three replicates of each treatment were used for the physicochemical analyses of the fruiting bodies. The mushrooms were weighed, dehydrated at 55 °C for approximately 36 hours, ground in a mini-processor, packed in polythene jars, sealed and stored under refrigeration at 5 ± 2 °C. The basidiocarps were analysed according to the procedures adopted by the A.O.A.C (1984) for moisture, lipid content and ash. Moisture was quantified in an oven at 105 °C for 6 hours. Crude fat content was determined by gravimetry after continuous extraction of the samples with sulphuric ether in Soxhlet equipment and ash was calculated by the mass of the sample after incineration. The protein fraction was determined using the "Kjeldahl" method, with the crude protein of the mushroom being determined from the nitrogen content, using the conversion factor N x 4.38 according to Miles and Chang (1997).

4.9.1. Carbohydrates and calorific value

They were estimated by difference from the moisture, fixed mineral residue, protein and lipids expressed in g/100g, using the following calculation (Equations 09 and 10) (INSTITUTO ADOLFO LUTZ, 1985):

$$\% \, Carboidratos = 100 - (A + B + C + D)$$

Eq. 09

In which:
A - % humidity
B - % fixed mineral residue
C - % protein
D - % lipids

Total calorific value: was calculated according to the calculation (ANVISA, 2003):

$$Valor\ Calórico\ Total\ (Kcal/100g) = (E \times 4) + (C \times 4) + (D \times 9)$$

Eq. 10

In which:
E - carbohydrates
C - proteins
D - lipids

4.9.2. Proteins Total

Total proteins were determined using the Kjedahl method according to the Manual of Food Analysis Practices (2020), which is based on wet combustion via heating with concentrated sulphuric acid in the presence of catalysts to reduce the organic nitrogen, ammonia, captured in an alkaline solution to form ammonium sulphate. The ammonia is then distilled off in a 4% boric acid solution followed by direct titration of the ammonia with a standard hydrochloric acid solution. The sample digestion procedure was carried out by carefully transferring 1.0 g of the samples to the Kjeldahl tube with 1.8 g of catalytic mixture and 10 mL of concentrated sulphuric acid without touching the tube walls. The tubes were placed in the digester at 350°C until the solution became colourless or slightly bluish. After cooling the tube, the walls were washed with 5ml of distilled water and then 3 drops of phenolphthalein were added. It was then attached to the distiller. The condenser outlet was immersed in an erlenmeyer flask with 25 mL of 4% boric acid HJBOJ solution. The boiler water was previously heated and then the distiller temperature was lowered to a scale of 3. The alkaline pH was obtained by adding 40% NaOh until the colour changed to purple. The distiller temperature was then increased to 10 in order to promote distillation. The boric acid solution containing the distilled protein was reserved for titration with 0.1M hydrochloric acid until a reddish colour appeared. The results were expressed as % nitrogen and the amount of protein was determined using Equation 11.

$$Proteínas\ totais(g/100g) = \frac{(Va - Vb) x\ fa\ x\ F\ x\ 0{,}14}{Peso\ da\ mostra}$$

Eq. 11

Va - volume of standardised 0.1N hydrochloric acid used to titrate the sample Vb - volume of standardised 0.1N hydrochloric acid used to titrate the blank
fa - nitrogen - vegetable protein matching factor: 5.75

4.9.3 Lipids Total

Total lipids were determined using the Bligh & Dyer method (1959) in which 2.5 g of the dried samples in triplicate were subjected to cold extraction of the total lipids in a mixture of chloroform, methanol and water. In this method, the lipids remain in the chloroform phase, which is then evaporated (Albuquerque and Sousa, 2024). The amount of lipids is obtained by weighing and the results are expressed in g per 100g of sample using Equation 12:

$$Lipídios\ totais\ (\%) = \frac{peso\ final - peso\ béquer\ x\ 4x\ 100}{Peso\ da\ mostra\ (g)}$$ Eq. 12

4.10. N, C and C/N content of substrates

4.10.1. Carbon organic

It was determined using the Walkley and Black method, modified from Tedesco et al. (1995). Each sample (0.01 g) was added to an erlenmeyer flask containing potassium dichromate (K2Cr2O7) and sulft'.ric acid (H_2 SO_4) and heated to 60°C. Two types of blank were prepared: one that was subjected to the same heating conditions and one that was not heated. After cooling, indicators were added and then the solution was titrated with ammoniacal ferrous sulphate (Fe (NH)$_{42}$ (SO)$_{42}$ .6H_2 O). The results were obtained from Equations 13 and 14:

$$A = \frac{((Vba - Vbn)(Vbn - Vba))}{Vbn + (Vba - Vam)}$$ Eq. 13

In which:
Vba: volume spent titrating the blank with heating; Vbn: volume spent titrating the blank without heating; Vam: volume spent titrating the sample.

$$Carbono\ (C)\% = \frac{(A)(molaridade\ do\ sulfato\ ferroso)(3)\ x\ 100}{(massa\ da\ amostra\ (mg))}$$ Eq. 14

In which:
3= ratio between the number of moles of dichromate reacting with iron, multiplied by the number of moles of dichromate reacting with carbon, multiplied by the atomic mass of carbon.

4.10.2. Nitrogen total

The Kjeldahl method was used to analyse total nitrogen, where the samples (0.1g) were digested in a catalyst mixture containing sulphuric acid ($H_2 SO_4$) in a Velp Scientifica digester block until a blue colour was obtained. After cooling, the reactions were distilled in a Velp Scientifica nitrogen distiller in a basic medium (NaOH 35%) and the product of this distillation was collected in an erlenmeyer flask in a 4% boric acid solution ($H_3 BO_3$) with indicators and titrated with 0.1 N hydrochloric acid (HCl) (Tedesco et al., 1995, Moretto et al., 2002a). The nitrogen concentration was determined according to Equation 15:

$$N(N)\% = \frac{V \, x \, Fc \, x \, 0{,}0014}{M} \quad \text{Eq. 15}$$

In which:
V = volume of HCl used in the titration, mL. Fc= HCl correction factor
M = sample mass (g)

4.10.3 C/N ratio

The C/N ratio was obtained from the proportion of carbon contained in each sample in relation to nitrogen.

4.11. Analysing statistics

The experimental design was entirely randomised complete blocks, using Tukey's test, with a significance level of $p<0.05$ (GOMES, 1990). Statistical analyses were carried out using the SAS (Statistical Analysis System) programme. The results were subjected to analysis of variance (ANOVA) followed by the Tukey test with a significance level of 5%.

5. RESULTS AND DISCUSSION

5.1. Vegetative and reproductive aspects of Pleurotus ostreatus var. florida

When the colony of P. ostreatus var. florida reached the edge of the Petri dish, it had a white cottony texture (Figure 16A). After 144 hours, it differentiated into a mycelium of septate hyaline hyphae, with obvious connecting staples and anastomoses (Figure 16B). This behaviour corroborates the literature, which describes the beginning of the life cycle of P. ostreatus with the germination of haploid basidiospores, progressing to a monokaryotic primary mycelium (Rodriguez et al., 2023). The anastomosis observed in our study facilitates the fusion of two compatible haploid hyphae, a critical process called plasmogamy. This fusion not only allows the exchange of cytoplasm and nutrients, but also nuclear transfer, which is essential for the formation of dikaryotic hyphae and the subsequent rapid expansion of the colony. This phenomenon is fundamental for the development of the secondary mycelium, the dominant phase in Basidiomycota, and for the formation of the tertiary mycelium, specialised in the creation of basidiomes (Barh et al., 2019; Figures 16A, 16C and 16D).

Figure 16. Vegetative and reproductive aspects of Pleurotus ostreatus var. florida

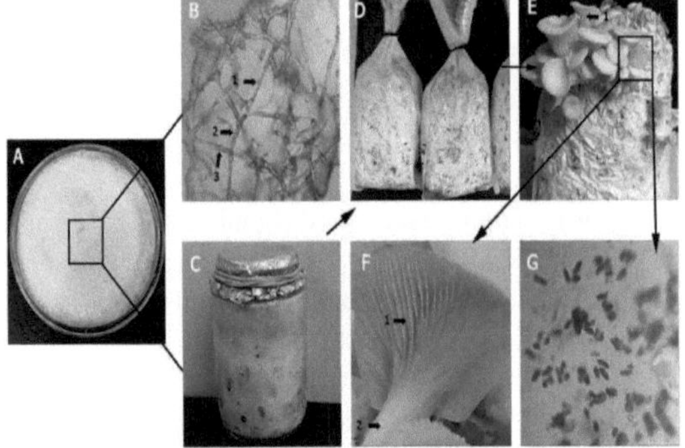

A: Colony of Pleurotus ostreatus var. florida with 15 days of growth in Sabouraud-dextrose agar culture medium); B: Vegetative aspect: 1 - connecting staple, 2 - hyphae and 3 - anastomosis; C: seed production on pre-cooked maize grain; D: Vegetative growth (green bean pods without the grain (80%) + malt bagasse (20%) in a production bioreactor 15 days after inoculation; E: Fruiting, E1 - Mushroom or hat or basidiome; F: Lamellar-shaped hymenium (F1), stem (F2); G: Sexual spores. Source: Authors, 2024.

The connecting clamps play a crucial role, ensuring nuclear division and the equitable distribution of nuclei between the daughter cells, thus maintaining the state dikaryotic mycelium. This was evidenced by the structuring of the lamellae in the basidiome, where karyogamy followed by meiosis occurs, culminating in the formation of four basidiospores per basidium (Figure 16F and 16G). When these spores germinate, they begin the cycle again by transforming into haploid mycelia which eventually become binucleate after the protoplasts fuse (Albuquerque; Sousa, 2021). Analysing these processes reinforces the descriptions already documented in the literature on P. ostreatus. Furthermore, it points to potential strategies for manipulating the phases of mycelial growth in controlled cultures. Thus, these interventions have the potential to optimise the commercial production of basidiomata, increasing the efficiency and quality of the fungi cultivated.

5.2. Physico-chemical characterisation of the co-products used

The physicochemical analyses of the substrates (Table 10) showed that malt bagasse (BM) contained 4.5% total sugars, 2.68% reducing sugars, 3.41% ash, 8.09% moisture, pH 5.94 and 1° Brix. The bean pod without grains (VF) contained 3.62% total sugars; 2.17% reducing sugars; 2.24% ash; 22.10% moisture; pH 6.30 and 1° Brix. The residual yeast contained 0.63% total sugars, 0.34% reducing sugars, 1.26% ash, 90.61% moisture, pH 5.10 and 9.8° Brix. The hot trub (TQ) contained 6.24% total sugars, 5.3% reducing sugars, 0.24% ash, 85.04% moisture, pH 5.28 and 13° Brix. Hop trub (TF) contained 5.25 per cent total sugars, 3.95 per cent reducing sugars, 1.2 per cent ash, 69.8 per cent moisture, pH 5.15 and 19° Brix. The VF + BM obtained a carbon content of 38.17 per cent and a nitrogen content of 2.27 per cent. The C/N ratio was 17:1.

Table 10. Physico-chemical composition of the substrates used.

Substrates	*Humidity (%)	AT (%)	ART(%)	ANR	Ash (%)	pH	°Brix
BM	8,09±0,14	4,50±0.011	2,68±0.009	1,82	3,41±0,04	5,94	1,0
VF	22,10±21,52	3,62±0,09	2,17±0,021	1,45	2,24±0,33	6,30	1,0
R	90,61±2,10	0,63±0,56	0,34±0,01	0,29	1,26±0,66	5,10	9,8
TQ	85,04±4,81	6,24±0,18	5,30±0,58	0,94	0,24±0,70	5,28	13,0
TF	69,80±0,79	5,25±0,59	3,95±0,40	1,30	1,2±0,10	5,15	19,0
CV(%)	8,79	10,55	6,96		12,01		

VF = Bean pods, BM= Malt bagasse, LR = Residual yeast, TQ= Hot trub, TF = Hop trub. AT = Total Sugars, ART = Total Reducing Sugars, ANR = Non-Reducing Sugars
* Moisture on a dry basis. Source: Authors, 2024.

The results indicate that the VF + BM combination has a carbon content of 38.17 per cent and a nitrogen content of 2.27 per cent, resulting in a C/N ratio of 17:1. This value is within the recommended range for best Pleurotus spp. production results, which varies between 7:1 and 40:1 (Elbagory et al., 2022; Feng et al., 2023). The right C/N ratio is crucial as it directly influences mycelium growth and mushroom fruiting. The pH of the substrates and supplements analysed is close to the ideal for growing Pleurotus spp. which is generally recommended between 4 and 6, being slightly acidic (Vilas et al., 2020). Keeping the pH within this range is essential to optimise the absorption of nutrients by the fungi and ensure an environment conducive to healthy growth.

The concentration of ash in the substrates and suitable supplements ensures good growth of P. ostreatus. An ash concentration of between 2.5% and 15.7% is recommended (Cueva et al., 2017), which suggests that the substrates used have an ash content suitable for growing this fungus and that the addition of supplements can contribute and improve the mineral content of the medium.

The substrates had different levels of reducing sugars and° Brix, however, all the formulations evaluated had a reducing sugar content greater than 50% of the total sugar content. Reducing sugars are important sources of energy for the initial growth of the mycelium and can also indirectly influence mushroom formation, while° Brix indicates the concentration of soluble solids in the substrate, reflecting its ability to provide nutrients available to the fungi. These data therefore suggest that these substrates have the potential to support a nutritionally rich environment for producing P. ostreatus var. florida, favouring not only biological efficiency but also the nutritional quality of the mushrooms produced.

5.3 Evaluation of the biological and production parameters of Pleurotus ostreatus var. florida

All the treatments showed continuous, abundant and vigorous growth (Figure 17), reaching complete colonisation in 16 days (Table 11). After opening the bioreactors, treatments T3 and T4 reached full development of the fruiting bodies in just 6 days, with a total cultivation cycle of 22 days. Then treatments T1 and T2 ended their cycles 7 days after opening the bioreactors, with a total colonisation time of 23 days (Figure 18). The combination of bean pods without grains and malt bagasse supplemented with liquid brewery co-products resulted in an adequate cultivation time for mushroom production of this species, with values close to those reported by (Dagnew, 2018; Maurya et al, 2019; De La Cruz-Marcos et al., 2023;) and superior in a difference of approximately 10 days to those reported by Hoa et al., (2015), Aisha. et al., (2019), Zakil et al., (2019), Dhakal et al., (2020), Muswati et al., (2021) and Tavarwisa et al., (2021) on different substrates.

Table 11. Biological parameters of Pleurotus ostreatus var. florida found in the first production flow.

Treatments	Spawn development	Moisture in formulations	Colonisation time (days)	Total cycle (days)	Stipe size (cm)	Stipe width (cm)	Pileus area (cm)2
T1	+++	75,9%	16	23	4,43±0,15	0,83±0,51	15,45±0,41
T2	+++	75,9%	16	23	3,80±0,05	0,96±0,57	11,33±0,65
T3	+++	75,9%	16	22	5,50±0,21	1,20±0,33	23,73±0,20
T4	+++	75,9%	16	22	4,40±0,19	1,16±0,69	15,19±0,19
(±)EP					2,97	5,08	8,69
CV (%)					1,30	3,07	5,73

+ sparse growth; ++ moderate growth; +++ abundant growth. CV: Coefficient of variation. SE: Standard error. T1: Green bean pods without grains (80 per cent) + Malt bagasse (20 per cent); T2: Green bean pods without grains (80 per cent) + Malt bagasse (20 per cent) + residual yeast (5 per cent); T3: Green bean pods without grains (80 per cent) + Malt bagasse (20 per cent) + residual yeast (5 per cent).grain (80 %) + malt bagasse (20 %) + hot trub (10 %); T4: green bean pods without grain (80 %) + malt bagasse (20 %) + hop trub (10 %). Source: Authors, 2024.

Figure 17. Substrates colonised by the primary mycelium of Pleurotus ostreatus var. florida after 16 days of growth in the dark, when the bioreactors were opened.

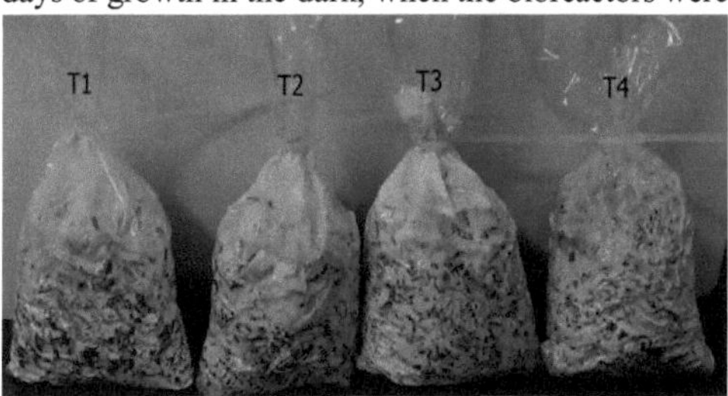

Source: Own authorship, 2024.

Figure 18. Macroscopic appearance of the fruiting bodies of Pleurotus ostreatus var. florida.

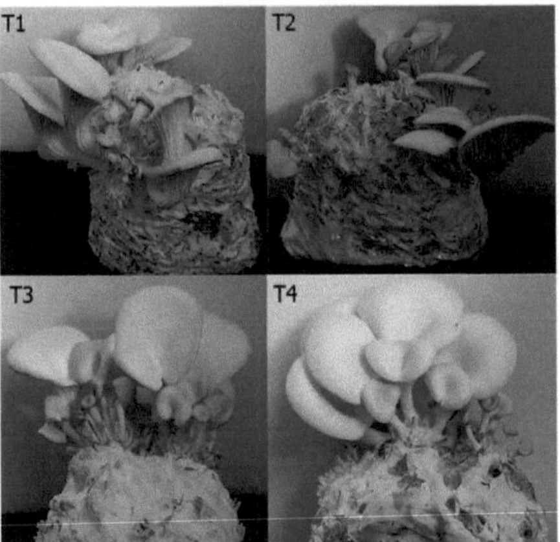

Source: Own authorship, 2024.

Extrinsic and intrinsic production factors can alter mycelial growth time and the formation of fruiting bodies (MUSWATI et al., 2021). The differences observed in relation to the literature can be explained mainly by the nutritional composition of the formulations. The substrates and supplements used in this work each have a nutritional composition suitable for growing P. ostreatus var. florida, with an appropriate concentration of carbon, nitrogen and ash. In addition, because the pH is slightly acidic and the sugar content is mainly in the reduced form, mycelial growth and the formation of fruiting bodies are accelerated, as the sugars are readily available for metabolic reactions and the enzymes used in the process act efficiently (Tavarwisa et al., 2021). Based on the characterisation of the fruiting body, treatment T3 had a height and width of 5.5 cm and 1.2 cm, respectively, and an area of 23.73 cm^2. Treatments T4 (4.4 cm height, 1.16 cm stipe width and 15.19 cm^2 area) and T1 (4.43 cm height, 0.83 cm stipe width and 61.7 cm^2 area) obtained similar results. Treatment T2 had a height of 3.8 cm, a width of 0.96 cm and an area of 11.33 cm^2.

These findings are in line with previous studies, which investigated similar parameters using different substrates (Kortei et al., 2018; Dubey et al., 2019; Sardar et al., 2020; Muswati et al., 2021; Wiafe-Kwagyan et al., 2022; Subedi et al., 2023; De La Cruz- Marcos et al., 2023). The morphometric dimensions of Pleurotus spp. basidiomata can vary significantly due to the nature of the substrate, air composition, luminosity, temperature and characteristics of the growing container. Nitrogen supplementation can increase crop productivity, but there is a limit, as high levels of nitrogen can inhibit the fruiting of Pleurotus spp. mushrooms (Bellettini et al., 2019). It should be noted that the brewery yeast used, despite its high protein content (up to 60 per cent of dry weight), has a low sugar concentration (0.63 per cent) (Table 10). The addition of yeast may have increased the nitrogen content in the substrate and unbalanced the C/N ratio, possibly harming fruiting and resulting in the smaller mushroom sizes observed in the corresponding treatments.Treatment T4 recorded the highest yield (13%) and biological efficiency (61.9%). Treatment T1 had a yield of 12.11% and biological efficiency of 40.37%, while T3 had a yield of 9.90% and biological efficiency of 33%. The lowest values were obtained by treatment T2, which showed a yield of 6.14% and biological efficiency of 14.33% (Table 12). With regard to productivity, the results in grams per day in descending order were as follows: T4 (1.54); T1(1.42); T3(1.42) and T2(1.04). There was a correlation between the values for biological efficiency and yield. This was because the total cultivation time for all the formulations was very close, with only 1 day of cultivation. for the T1 and T2 formulations. However, this difference of 1 day was enough for treatment T3 to obtain the same productivity values as treatment T1, even though it had a lower biological efficiency.Higher yield values are associated with higher biological efficiency values. This is because both take into account the fresh weight of the mushroom, the difference lies in the humidity of the substrate (Muswati et al., 2021). This explains the positive correlation between the results obtained for these two variables. With regard to the individual performance of each formulation, extrinsic and intrinsic factors must be taken into account, as well as the influence of biological parameters. The weight of the mushroom (g/bioreactor) depends on the diameter of the pileus, the length and thickness of the stipe, the number of effective fruiting bodies per bioreactor and the thickness of the pileus (Wiafe-Kwagyan et al., 2022). Thus, as lower biological efficiency values are related to lower values for stipe size and pileus diameter, the lower values for morphometric dimensions obtained for treatment T2 partly explain the low yield and biological efficiency of this formulation. In addition, the lower moisture content in the mushrooms produced from the T2 formulation resulted in a lower mass and, consequently, lower biological efficiency and yield values. In this context, Sardar et al. (2020) obtained results close to those of treatment T2 when evaluating formulations based on cotton by-products and (Tavarwisa et al., 2021) reported slightly higher results when evaluating formulations based on wheat, maize and sawdust. The results obtained in treatments T1 and T3 were close to those observed by Fufa et al. (2021) when they used corn cobs, millet straw and bamboo in different formulations. With regard to treatment T4, similar results were obtained when using sisal-based formulations (Carmo et al., 2021) and mixtures of palm kernel, sugarcane bagasse and rice straw (Zakil et al., 2019).

The loss of organic matter (LOM) between treatments ranged from 49.25% to 52.66% and compost consumption (CC) ranged from 47.33% to 51.00%, with no statistical difference between treatments. T3 had the highest PMO (51.48%) and, consequently, the lowest CC (49.52). T4 had an MOP of 49% and a CC of 51%.

Despite being complementary values, in practice these results have different meanings and help growers visualise the performance of the crop from different perspectives. A lower PMO represents a greater chance of the formulation achieving a second harvest flow, as the substrate still has the organic matter needed for the fungus to develop. As for CC, a lower value is associated with the fungus' greater ability to absorb nutrients from the medium and develop. According to the results obtained, if the fungus consumed around 50 per cent of the substrates in the first cultivation, all the formulations are capable of generating a second production flow.

Table 12. Characterisation of the product parameters of Pleurotus ostreatus var. florida

Treatments	Mushroom moisture (%)	Yield (%)	Biological efficiency (%)	Productivity (g/day)	Loss of organic matter (%)	Compost consumption (%)
T1	69,99±0,12	12,11±1,02	40,37±0,65	1,42±1,36	49,25±0,69	50,74±0,48
T2	44,18±0,78	6,14±0,09	14,33±0,29	1,04±0,49	51,48±0,15	49,52±0,11
T3	68,38±0,98	9,90±0,50	33,00±1,70	1,42±0,05	52,66±0,60	47,33±0,09
T4	73,84±0,25	13,00±0,35	61,90±1,29	1,54±0,59	49,00±0,08	51,00±0,85
Standard deviation	5,97	11,94	0,99	0,09	4,97	7,52
CV (%)§	10,50	27,61	18,02	20,88	8,51	14,77

CV. Coefficient of variance**. T1**: Green bean pods without grains (80%) + malt bagasse (20%); **T2**: Green bean pods without grains (80%) + malt bagasse (20%) + residual yeast (5%); **T3**: Green bean pods without grains (80%) + malt bagasse (20%) + hot trub (10%); **T4**: Green bean pods without grains (80%) + malt bagasse (20%) + hop trub (10%). Source: Authors, 2024.

5.4 Nutritional aspects of Pleurotus ostreatus var. florida

Based on the formulations tested, the basidiomata formed had the following characteristics: T1 had a carbohydrate content of 56.82%, protein of 27.24%, ash of 4.24%, lipids of 3.89% and a calorific value of 417.80 Kcal/100g. T2 had a carbohydrate content of 62.72%, protein of 25.81%, ash of 3.74%, lipids of 2.82% and a calorific value of 423.56 Kcal/100g. T3 had a carbohydrate content of 56.48%, protein of 28.82%, ash of 3.95%, lipids of 3.52% and a calorific value of 422.02 Kcal/100g. Finally, T4 had a carbohydrate content of 63.18%, protein 23.22%, ash 4.01%, lipids 3.62% and a calorific value of 417.87 Kcal/100g (Table 13).

Table 13. Physicochemical composition of Pleurotus ostreatus basidiomes.

Carbohydrates		Proteins	Lipids	Ashes	Calories
Treatments	(%)	(%)	(%)	(%)	(Kcal/ 100g)
T1	56,82±0,61	27,24±0,09	3,89±0,49	4,24±0,39	417,80±0,18
T2	62,72±0,28	25,81±0,75	2,82±0,25	3,74±0,08	423,56±0,36
T3	56,48±0,40	28,82±0,32	3,52±0,48	3,95±0,53	422,02±0,19
T4	63,18±0,18	23,22±0,65	3,62±0.29	4,01±0,20	417,87±0,28
CV (%)	0,75	2,33	4,83	3,03	1,73

CV. Coefficient of variance**. T1**: Green bean pods without grains (80%) + malt bagasse (20%); **T2**: Green bean pods without grains (80%) + malt bagasse (20%) + residual yeast (5%); **T3**: Green bean pods without grains (80%) + malt bagasse (20%) + hot trub (10%); **T4**: Green bean pods without grains (80%) + malt bagasse (20%) + hop trub (10%). Source: Authors, 2024.

In this study, all the formulations tested produced mushrooms with a high protein content. Formulation T2 had a low lipid content (maximum 3g of total fat in 100g) in accordance with RDC Nº 54/2012, which regulates the Mercosur Complementary Nutritional Information. In addition, all the mushrooms produced showed an ash content above the standardised value in the TACO Table of 1.8%, indicating a food rich in minerals and nutritionally safe. The greater amount of nitrogen in the substrate has a positive correlation with the protein content of the basidiome, as observed by Raman et al. (2021). Due to its importance in protein synthesis, Koutrotsios et al. (2019), when investigating the nutritional composition of P. ostreatus grown on olive coproduct substrates, noted that the increase in the mushrooms' ash and protein contents was associated with higher concentrations of minerals and nitrogen in the substrate. In addition, a lower C:N ratio benefits the production of crude protein, amino acids and nucleotides at high levels. Of the supplements evaluated, hot trub (T3) was the only one that significantly increased the mushroom's protein content when compared to the formulation without supplementation (T1). This result can be attributed to the increased nitrogen content in the substrate, since the proteins in the trub make up around 40 to 70 per cent of its dry weight (Jaeger et al., 2020). In addition, the morphometric dimensions of the mushroom can also be attributed to the increased nitrogen content in the substrate. may have contributed, since treatment T3 had the largest pileus area, which is the part of the mushroom with the highest protein concentration (JOSIANE et al., 2018). The addition of yeast (T2) should theoretically increase the protein content of the mushroom for the same reason as the hot trub. However, the observed reduction in mushroom dimensions in this treatment may have counterbalanced the expected effects, resulting in a lower protein content than anticipated. This suggests that although yeast is a rich source of nitrogen, other morphometric and physiological factors of the mushroom can significantly influence protein accumulation. With regard to the hop trub (T4), the dimensions of the mushroom were similar to those of the T1 treatment, but this formulation had the lowest protein content. This reduction can be explained by the specific composition of hop trub, which may have a lower protein content and a different nutrient profile compared to hot trub. In addition, the possible presence of growth inhibitors or compounds that hinder the efficient absorption of nutrients by the mushrooms may have contributed to these results. These factors can alter the fungus's metabolism, leading to less incorporation of proteins and other essential nutrients. Studies into the composition of hop trub are therefore needed to understand in more detail how it affects the nutritional content of the mushrooms. All the supplemented formulations showed a reduction in the concentration of lipids and ash when compared to the formulation without supplementation. This result highlights the complexity involved in the nutritional composition of mushrooms. Although higher protein levels are expected to be related to lower lipid levels, the results for formulations T2 and T4 show that the nutritional composition of mushrooms can be influenced in different ways by the components present in the substrate. The variation in composition indicates that supplementation can alter nutritional profiles in a non-linear way, impacting the concentration of various nutrients in a complex and interdependent manner. Therefore, the use of different types of co-products as substrates should be carefully evaluated to better understand their effects on the final nutritional composition of the mushrooms. In addition, the average protein, carbohydrate and lipid content of the mushrooms produced in this study was higher than the average of 48 other formulations (Table 01) used to produce P. ostreatus, where the variation in composition ranged from 10.09% to 30.46% for proteins, 2.11% to 65.9% for carbohydrates and 1.2% to 7.23% for lipids. And which include cotton, oil palm, rice and wheat, sugar cane, sisal, bean pods, maize, bananas and cassava (Figure 19). This highlights the nutritional potential of mushrooms grown in substrates supplemented with brewery and bean co-products.Individually, the formulations showed promise for generating mushrooms with a high protein content. Specifically, the formulation with hot trub (T3) proved to be the most efficient, which can be exploited to optimise the production of mushrooms with high nutritional value. These results suggest that using by-products from the brewing industry, particularly hot trub, not only contributes to environmental sustainability by recycling by-products, but also improves the nutritional quality of the mushrooms produced.

Figure 19. Comparison of the average nutritional composition of 52 formulations.

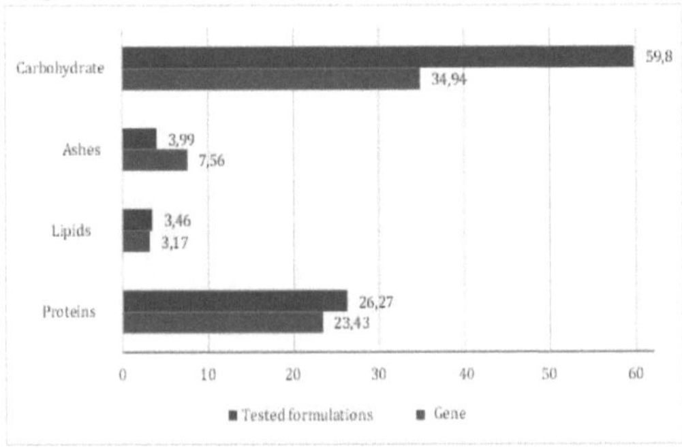

Source: author

6. CONCLUSIONS

Based on the results obtained under the conditions established and evaluated in this study, it can be concluded that:

• The individual substrates showed satisfactory characteristics for the composition of formulations for the production of P. ostreatus var. florida in terms of pH, sugars, ash and soluble solids, favouring fungal growth and fruiting;

• Hop trub (T4) was the most effective supplement for increasing biological efficiency compared to the T1 treatment without supplementation.

• The formulations tested are within the optimum range for producing mushrooms with a higher yield and protein and mineral content, according to RDC No. 54/2012.

• The mushrooms had an ash content significantly above the value standardised by the TACO table, indicating that they are mineral-rich and nutritionally safe foods.

• Hot trub (T3) was the most effective as a supplement because it generated mushrooms with the highest pileus diameter values and ensured a protein increase in the mushrooms compared to the formulation without supplementation (T1).

REFERENCES BIBLIOGRAPHIC

AGHAJANI, H. et al. Influence of relative humidity and temperature on cultivation of Pleurotus species. Wood. Ciencia y tecnología, v. 20, n. 4, p. 571-578, 2018. http://dx.doi.org/10.4067/S0718-221X2018005004501

AGUIAR, Dayana; PEREIRA, Ana C.; MARQUES, José C. The influence of transport and storage conditions on beer stability-a systematic review. Food and Bioprocess Technology, v. 15, n. 7, p. 1477-1494, 2022. https://doi.org/10.1007/s11947-022-02790-8

AISHA, E.; HASSAN, S.; SH, A.-. Evaluation of oyster mushroom (Pleurotus ostreatus) cultivation using different organic substrates. Alexandria Science Exchange Journal, v. 40, p. 427-440, 2019. https://doi.org/10.21608/asejaiqjsae.2019.49370

ALBUQUERQUE, V.; SOUSA, A. C. B. Axenic cultivation of Pleurotus ostreatus var. florida.1.ed. Editoria Novas Edições Acadêmicas, p. 7-28; 2024.

ALBUQUERQUE, V.; SOUSA, A. C. B. Fungiculture of the edible mushroom Pleurotus ostreatus var. florida. 1.ed. Editoria Novas Edições Acadêmicas, p. 69; 2021.

ALSANAD, M. A. et al. Spent coffee grounds influence on Pleurotus ostreatus production, composition, fatty acid profile, and lignocellulose biodegradation capacity. CyTA-Journal of Food, v. 19, n. 1, p. 11-20, 2021.

AMORIELLO, T.; CICCORITTI, R. Sustainability: Recovery and Reuse of Brewing-Derived By-Products. Sustainability, v. 13, p. 2355, 2021. https://doi.org/10.3390/su13042355

AMORIELLO, Tiziana et al. Technological properties and consumer acceptability of bakery products enriched with brewers' spent grains. Foods, v. 9, n. 10, p. 1492, 2020. https://doi.org/10.3390/foods9101492

AN, Q.; LI, C. S.; YANG, J.; CHEN, S. Y.; MA, K.-Y. et al. Evaluation of laccase production by two white-rot fungi using solid-state fermentation with different agricultural and forestry residues. BioResources, v. 16, n. 3, 2021.

ARAÚJO, N. L. et al. Production of mycelial biomass and lignocellulolytic enzymes of Pleurotus spp. in liquid culture medium. Research, Society and Development, v. 10, n. 1, p. e6810111406-e6810111406, 2021. https://doi.org/10.33448/rsd-v10i1.11406

ASSOCIATION OF OFFICIAL ANALYTICAL CHEMISTS (AOAC). Official methods of analysis. 16th ed. Washington, 1995.

ATTARAN DOWOM, S.; REZAEIAN, S.; POURIANFAR, H. R. Agronomic and environmental factors affecting cultivation of the winter mushroom or Enokitake: Achievements and prospects. Applied Microbiology and Biotechnology, v. 103, n. 6, p. 2469-2481, 2019. https://doi.org/10.1007/s00253-019-09652-y

ATLI, Burcu et al. Solid state fermentation optimization of Pleurotus ostreatus for lovastatin production. Pharmaceutical Chemistry Journal, v. 53, p. 858-864, 2019. https://doi.org/10.1007/s11094-019-02090-0

BARH, A.; SHARMA, V.P.; ANNEPU, S.K.; KAMAL, S.; SHARMA, S.; BHATT, P. Genetic improvement in Pleurotus (oyster mushroom): a review. 3 Biotech, v. 9, p. 322, 2019. https://doi.org/10.1007/s13205-019-1854-x

BARBOSA, F. R. et al. Beans resistant to golden mosaic. Santo Antônio de Goiás: Embrapa Rice and Beans, 2021.

BAYRAM, Ö. S.; BAYRAM, Ö. An anatomy of fungal eye: fungal photoreceptors and signalling mechanisms. Journal of Fungi, v. 9, n. 5, p. 591, 2023. https://doi.org/10.3390/jof9050591

BELLETTINI, M.B. et al. Factors affecting mushroom Pleurotus spp. Saudi Journal of Biological Sciences, v. 26, p. 633-646, 2019. https://doi.org/10.1016/j.sjbs.2016.12.005

BLIGH, E. G.; DYER, W. J. A rapid method of total lipid extraction and purification. Can. J. Biochem. Physiol, v. 37, n. 8, p. 911-917, 1953.

BONATTI, M.; KARNOPP, P.; SOARES, H. M.; FURLAN, S. A. Evaluation of Pleurotus ostreatus and Pleurotus sajor-caju nutritional characteristics when cultivated in different lignocellulosic wastes. Food Chem., 88(3): 425-428, 2004.
BRAZIL. Ministry of Health. Resolution RDC no. 360, of 23 December 2003. Technical Regulation on nutritional labelling of packaged foods. ANVISA - National Health Surveillance Agency, Virtual Health Library: Brazil, DF, 2003.

CUEVA, M. B. R.; HERNÁNDEZ, A.; NIÑO-RUIZ, Z. Influence of C/N ratio on productivity and the protein contents of Pleurotus ostreatus grown in different residue mixtures. Revista de la Facultad de Ciencias Agrarias, v. 49, n. 2, p. 331-344, 2017.

DAGNEW. Substrate optimization for cultivation of Pleurotus ostreatus on lignocellulosic wastes (coffee, sawdust, and sugarcane bagasse) in Mizan-Tepi University, Tepi Campus, Tepi Town. Journal of Applied Biology and Biotechnology, p. 14-20, 2018. https://doi.org/10.7324/JABB.2018.60403

DAUD, M. et al. Habitat characteristics and utilisation of edible wild mushrooms by local communities in the protected forest in Pinrang Regency, Indonesia. In: IOP Conference Series: Earth and Environmental Science. IOP Publishing, 2021. p. 012125.

DE LA CRUZ-MARCOS, R.N. et al. Possible effects of different types of agricultural wastes on food security and mushroom (Pleurotus ostreatus) production. Brazilian Journal of Biology, v. 83, e273829, 2023. https://doi.org/10.1590/1519-6984.273829

DE MASTRO, F.; TRAVERSA, A.; MATARRESE, F.; COCOZZA, C.; BRUNETTI, G. Influence of Growing Substrate Preparation on the Biological Efficiency of Pleurotus ostreatus. Horticulturae 2023, 9, 439. https://doi.org/10.3390/horticulturae9040439

DHAKAL, P. et al. Growth and yield performance of oyster mushroom (Pleurotus ostreatus) on different substrates. Agriculture World, v. 8, p. 1-8, 2020. https://doi.org/10.38112/agw.2020.v08i01.001

DOROSKI, A. et al. Food waste originated material as an alternative substrate used for the cultivation of oyster mushroom (Pleurotus ostreatus): a review. Sustainability, v. 14, n. 19, p. 12509, 2022. https://doi.org/10.3390/su141912509

DO CARMO, C. O. et al. Bioconversion of sisal agro-industrial waste into high protein oyster mushrooms. Bioresource Technology Reports, v. 14, p. 100657, 2021.

https://doi.org/10.1016/j.biteb.2021.100657

DUBEY, D. et al. Comparative study on effect of different substrates on yield performance of oyster mushroom. Global Journal of Biology, Agriculture, Health Sciences, v. 7, 2019. 10.24105/2319-

DURÁN-ARANGUREN, D.D. et al. Biological pretreatment of fruit residues using the genus Pleurotus: A review. Bioresource Technology Reports, v. 16, 100849, 2021. https://doi.org/10.1016/j.biteb.2021.100849

EFFIONG, M. E. et al. Assessing the nutritional quality of Pleurotus ostreatus (oyster mushroom). Frontiers in Nutrition, v. 10, p. 1279208, 2024. https://doi.org/10.3389/fnut.2023.1279208

ELKANAH, F.A.; OKE, M.A.; ADEBAYO, E.A. Substrate composition effect on the nutritional quality of Pleurotus ostreatus (MK751847) fruiting body. Heliyon, v. 8, e11841, 2022. https://doi.org/10.1016/j.heliyon.2022.e11841

EL-RAMADY, H. et al. Green Biotechnology of Oyster Mushroom (Pleurotus ostreatus L.): A Sustainable Strategy for Myco-Remediation and Bio-Fermentation. Sustainability, v. 14, 3667, 2022. https://doi.org/10.3390/su14063667

EWEKEYE, T. S. et al. Effect of different substrates on the nutritional composition of Pleurotus ostreatus (Jacq.) P. Kumm.(Oyster Mushroom). Asian Journal of Research in Botany, v. 4, n. 4, p. 100-105, 2020.

FUFA, B.K.; TADESSE, B.A.; TULU, M.M. Cultivation of Pleurotus ostreatus on Agricultural Wastes and Their Combination. International Journal of Agronomy, v. 2021, p. 1-6, 2021. https://doi.org/10.1155/2021/1465597

FURLAN, S. A.; VIRMOND, L. J.; MIERS, D.; BONATTI, M.; GERN, R. M. M.; JONAS, R. Mushroom strains able to grow at high temperatures and low pH values. World Journal of Microbiology and Biotechnology, v. 13, p. 689-692, 1997.

GALLOTTI, F.; LAVELLI, V. The effect of UV irradiation on vitamin D2 content and antioxidant and antiglycation activities of mushrooms. Foods, v. 9, n. 8, p. 1087, 2020. https://doi.org/10.3390/foods9081087

GARUBA, T. et al. Influence of substrates on the nutritional quality of Pleurotus pulmonarius and Pleurotus ostreatus. Ceylon Journal of Science, v. 46, p. 67, 2017. https://doi.org/10.4038/cjs.v46i1.7419

GODSWILL, Awuchi Godswill et al. Health benefits of micronutrients (vitamins and minerals) and their associated deficiency diseases: A systematic review. International Journal of Food Sciences, v. 3, n. 1, p. 1-32, 2020. https://doi.org/10.47604/ijf.1024

GOMES, F. P. Curso de Estatística Experimental. Piracicaba: Nobel, ed. 13, 1990, p. 476, 1990.

GONZÁLEZ, Abigail et al. Evaluation of functional and nutritional potential of a protein concentrate from Pleurotus ostreatus mushroom. Food Chemistry, v. 346, p. 128884, 2021. https://doi.org/10.1016/j.foodchem.2020.128884

HOA, H.T.; WANG, C.-L.; WANG, C.-H. The Effects of Different Substrates on the Growth, Yield, and Nutritional Composition of Two Oyster Mushrooms (Pleurotus ostreatus and Pleurotus cystidiosus).Mycobiology,v.43,p.423-434,2015.

https://doi.org/10.5941/MYCO.2015.43.4.423

HOLTZ, M. Utilisation of cotton waste from the textile industry for the production of fruiting bodies of Pleurotus ostreatus DSM 1833. Journal of Environmental Sciences. Vol. 3, p. 37-51, 2009.

HUSSEIN, Ahmed M.; ALSHUKRI, Aqeel Y.; MOHAMMED, Akeel E. Susceptibility of Pleurotus ostreatus to grow on wheat, water hyacinth, barley straw and biodegrade its residues. International Journal of Agricultural and Statistical Sciences, v. 15, n. 2, p. 585-589, 2019.

HU, Y. et al. Effects and Mechanism of the Mycelial Culture Temperature on the Growth and Development of Pleurotus ostreatus (Jacq.) P. Kumm. Horticulturae, v. 9, n. 1, p. 95, 2023. https://doi.org/10.3390/horticulturae9010095

ADOLFO LUTZ INSTITUTE. Analytical standards, chemical and physical methods for food analysis. Ed. 3, São Paulo: Instituto Adolfo Lutz, 1985.

IRYNA, B. et al. Effect of perforation size and substrate bag fruiting position on the morphology of fruiting bodies and clusters in Pleurotus ostreatus (Jacq.) P. Kumm. Journal of Applied Biology and Biotechnology. https://doi.org/10.7324/JABB.2021.9305

JAEGER, A. et al. Brewer's spent yeast (BSY), an underutilised brewing by-product. Fermentation, v. 6, n. 4, p. 123, 2020. https://doi.org/10.3390/fermentation6040123

JOSIANE, M. et al. Effect of substrates on nutritional composition and functional properties of Pleurotus ostreatus. Current Research in Agricultural Sciences, v. 5, n. 1, p. 15-22, 2018.

JONGMAN, M.; KHARE, K. B.; LOETO, Daniel. Oyster mushroom cultivation at different production systems: A. European journal of biomedical and pharmaceutical sciences, v. 5, n. 5, p. 72-79, 2018.

KAUR, M. et al. Influence of artificial light on vegetative growth of Pleurotus eous and Pleurotus florida: An oyster mushroom. 2022.

KALILI, A.; OUAFI, R.; NACIRI, K. Chemical composition and antioxidant activity o f extracts from Moroccan fresh fava beans pods (Vicia faba L.). Baza Agro, v. 73, n. 1, 2022.

KARLOVIé, A. et al. By-products in the malting and brewing industries-re-usage possibilities. Fermentation, v. 6, n. 3, p. 82, 2020.
https://doi. org/10.3390/fermentation6030082

KORTEI, N. K. et al. Correlations of cap diameter (pileus width), stipe length and biological efficiency of Pleurotus ostreatus (Ex. Fr.) Kummer cultivated on gamma-irradiated and steam-sterilised composted sawdust as an index of quality for pricing. Agriculture & Food Security, v. 7, p. 1-8, 2018. https://doi.org/10.1186/s40066-018-0185-1

KOUTROTSIOS, G. et al. Elemental content in Pleurotus ostreatus and Cyclocybe cylindracea mushrooms: Correlations with concentrations in cultivation substrates and effects on the production process. Molecules, v. 25, n. 9, p. 2179, 2020. https://doi.org/10.3390/molecules25092179

LESA, K. N. et al. Nutritional Value, Medicinal Importance, and Health-Promoting Effects of Dietary Mushroom (Pleurotus ostreatus). Journal of Food Quality, v. 2022, n. 1, p. 2454180, 2022. https://doi.org/10.1155/2022/2454180

LI, Y. et al. Sexual spores in edible mushroom: bioactive components, discharge mechanisms and effects on fruiting bodies quality. Food Science and Human Wellness, v. 12, p. 2111-2123, 2023. https://doi.org/10.1016/j.fshw.2023.03.014

LYNCH, K. M.; STEFFEN, E. J.; ARENDT, E. K. Brewers' spent grain: a review with an emphasis on food and health. Journal of the Institute of Brewing, v. 122, n. 4, p. 553-568, 2016. https://doi.org/10.1002/jib.363

MARINHO, G.C.S.; SOUSA, A.C.B. Cultivation of black shimeji-in agro-industrial waste. 1.ed. Editoria Novas Edições Acadêmicas, p. 7-28; 2024.

MARTÍNEZ-FLORES, H.E.; CONTRERAS-CHÁVEZ, R.; GARNICA-ROMO, Ma.G. Effect of Extraction Processes on Bioactive Compounds from Pleurotus ostreatus and Pleurotus djamor: Their Applications in the Synthesis of Silver Nanoparticles. Journal of Inorganic and Organometallic Polymers, v. 31, p. 1406-1418, 2021. https://doi.org/10.1007/s10904-020-01820-2

MAURYA, A. et al. Effect of media and substrates for spawn production of dhingri mushroom (Pleurotus ostreatus). Journal of Natural Resource and Development, v. 14, n. 2, p. 88-92, 2019. https://doi.org/10.13140/RG.2.2.21808.79362

MBASSI JOSIANE, E.G. et al. Effect of Substrates on Nutritional Composition and Functional Properties of Pleurotus ostreatus. Ceylon Journal of Science, v. 5, p. 15-22, 2018. https://doi.org/10.18488/journal.68.2018.51.15.22

MILES, P. G.; CHANG, S. T. Biologia de las setas: fundamentos básicos y acontecimientos actuales. World Scientific, Hong Kong, 1997.

MORETTO, E; FETT, R; GONZAGA, L.V; KUSKOSKI, E.M. 2002a. Composition centesimal of plants and other materials. 2nd ed. UFRGS, 174 p. Food products. Chapter 1. in: Moretto et al. (eds). Introduction to food science.

MUSWATI, C. et al. The Effects of Different Substrate Combinations on Growth and Yield of Oyster Mushroom (Pleurotus ostreatus). International Journal of Agronomy, v. 2021, p. 1-10, 2021. https://doi.org/10.1155/2021/9962285

NAVARRO, M.; GEA, F.; GIMENEZ, A.; MARTINEZ, A.; RAZ, D.; LEVANON, D.;

OFFER, D. Agronomical valuation of a drip irrigation system in a commercial mushroom farm. Scientia Horticulturae,v.265,ed.30,2020. https://doi.org/10.1016/j.scienta.2020.109234

NWOKO, M. C. et al. Variability effect of pH on yield optimisation and mycochemical compositions of Pleurotus ostreatus sporophores cultured on HCl-induced substrate. African Journal of Plant Science, v. 16, n. 5, p. 97-110, 2022. https://doi.org/10.5897/AJPS2021.2162

ODUNMBAKU, O. K.; ADENIPEKUN, C. O. Cultivation of Pleurotus ostreatus (Jacq Fr.) Kumm on Gossypium hirsutum Roxb. (Cotton waste) and Gmelina arborea L. sawdust. International Food Research Journal, v. 25, n. 3, 2018.

OBIAIGWE, J.A. et al. Growth, Yield and Nutritional Quality of Pleurotus pulmonarius and Pleurotus ostreatus, Grown on Different Substrates Amended with Wheat Bran. Biotechnology Journal International, v. 27, p. 46-60, 2023. https://doi.org/10.9734/bji/2023/v27i4690

OLAJIRE, A. A. The brewing industry and environmental challenges. Journal of cleaner production, v. 256, p. 102817, 2020. https://doi.org/10.1016/j.jclepro.2012.03.003

OLIVEIRA DO CARMO, C. et al. Bioconversion of sisal agro-industrial waste into high protein oyster mushrooms. Bioresource Technology Reports, v. 14, 100657, 2021. https://doi.org/10.1016/j.biteb.2021.100657

OMURA, M ., FURLAN, F ., PELLIZZARO, V . , FREIRIA, G ., SOUSA, W . , & TAKAHASHI, L. (2020). Application times and doses of nitrogen fertiliser for beans. Revista Terra & Cultura: Cadernos De Ensino E Pesquisa, 36(71), 124-136

PETER, O. E. et al. Utilisation of some agro-wastes for cultivation of Pleurotus ostreatus (Oyster Mushroom) in Keffi Nigeria. Environmental Microbiology, v. 5, n. 2, p. 60-69, 2019. https://doi.org/10.11648/j.fem.20190502.13

RACHWAL, K. et al. Utilisation of brewery wastes in food industry. PeerJ, v. 8, p. e9427, 2020. https://doi.org/10.7717/peerj.9427

RAJARATHNAM, S.; BANO, Z.; MILES, P. G. Pleurotus mushrooms. Part I A. morphology, life cycle, taxonomy, breeding, and cultivation. Critical Reviews in Food Science & Nutrition, v. 26, n. 2, p. 157-223, 1987.

RAKIB, M.R.M.; LEE, A.M.L.; TAN, S.Y. Corn husk as lignocellulosic agricultural waste for the cultivation of Pleurotus florida mushroom. BioResources, v. 15, p. 7980-7991, 2020. https://doi.org/10.15376/biores.15.4.7980-7991

RAMAN, J. et al. Cultivation and Nutritional Value of Prominent Pleurotus spp.: An Overview. Mycobiology, v. 49, p. 1-14, 2021. https://doi.org/10.1080/12298093.2020.1835142

RODRIGUEZ, L.M. et al. Protein recovery from brewery solid wastes. Food Chemistry, v. 407, 134810, 2023. https://doi.org/10.1016/j.foodchem.2022.134810

SARAIVA, B. R. et al. Waste from brewing (trub) as a source of protein for the food industry. International Journal of Food Science & Technology, v. 54, n. 4, p. 1247-1255, 2019. https://doi.org/10.1111/ijfs.14101

SARDAR, A.; SATANKAR, V.; JAGAJANANTHA, P.; MAGESHWARAN, V. Effect of substrates (cotton stalks and cotton seed hulls) on growth, yield and nutritional composition of two oyster mushrooms (Pleurotus ostreatus and Pleurotus florida). 2020.

SAPUTERA, A. et al. Effect of Watering Frequency on the Growth and Yield of Oyster Mushrooms (Pleurotus ostreatus). AGROSAINSTEK: Jurnal Ilmu dan Teknologi Pertanian, v. 4, n. 2, p. 155-160, 2020.

SENNA FERREIRA COSTA, F. et al. Reuse of hot trub as an active ingredient with antioxidant and antimicrobial potential. Waste and Biomass Valorisation, v. 12, p. 2037-2047, 2021.

SHAHBANDEH. Global production of mushrooms and truffles 2012-2022. 2024. Statista. Available at: https://www.statista.com/statistics/1018488/global-mushrooms-and-truffles-production/. Accessed on: 18 Sep. 2024.

SOH, E. et al. Effect of common foods as supplements for the mycelium growth of Ganoderma lucidum and Pleurotus ostreatus on solid substrates. PLoS One, v. 16, n. 11, p. e0260170, 2021. https://doi.org/10.1371/journal.pone.0260170

SUBEDI, S.; KUNWAR, N.; PANDEY, K.R.; JOSHI, Y.R. Performance of oyster mushroom (Pleurotus ostreatus) on paddy straw, water hyacinth and their combinations. Heliyon, v. 9, e19051, 2023. https://doi.org/10.1016/j.heliyon.2023.e19051

SUN, X; DOU, Z; SHURSON, G; HU, B. Fungal bioprocessing of wheat straw with fruit and vegetable discards to produce cattle feeds for enhanced sustainability. Resources, Conservation and Recycling. Vol. 199, 2023. https://doi.org/10.1016/j.resconrec.2023.107251

SYDOR, M.; COFTA, G.; DOCZEKALSKA, B.; BONENBERG, A. Fungi in Mycelium-Based Composites: Usage and Recommendations. Materials, v. 15, 6283, 2022. https://doi.org/10.3390/ma15186283

TAVARWISA, D.M.; GOVERA, C.; MUTETWA, M.; NGEZIMANA, W. Evaluating the Suitability of Baobab Fruit Shells as Substrate for Growing Oyster Mushroom (Pleurotus ostreatus). International Journal of Agronomy, v. 2021, p. 1-7, 2021. https://doi.org/10.1155/2021/6620686

TEDESCO, M.J; GIANELLO, C; BISSANI, C.A; BOHNEN, H; VOLKWEISS, S.J. 1995. Soil analyses, UFSC, pg. 19-56.

TOROS, G.; EL-RAMADY, H.; PROKISCH, J. Edible Mushroom of Pleurotus spp.: A Case Study of Oyster Mushroom (Pleurotus ostreatus L.). Environment, Biodiversity and Soil Security, v. 6, p. 51-59, 2022. https://doi.org/10.21608/jenvbs.2022.117554.1161

VAZ JUNIOR, S. Utilisation of agro-industrial waste: a sustainable approach. Brasília, DF:

Embrapa Agroenergia, 2020. 29 p. (Embrapa Agroenergia / Documentos, 31).

VILAS, P. M. et al. Effect of cultural variability on mycellial growth of eleven mushroom isolates of Pleurotus spp. Journal of Pharmacognosy and Phytochemistry, v. 9, n. 6, p. 881-888, 2020.

VIVIAN, A. F. et al. Mass spectrometry for the characterisation of brewing process. Food Research International, v. 89, p. 281-288, 2016. https://doi.org/10.1016/j.foodres.2016.08.008

WAN MAHARI, W.A. et al. A review on valorisation of oyster mushroom and waste generated in the mushroom cultivation industry. Journal of Hazardous Materials, v. 400, 123156, 2020. https://doi.org/10.1016/j.jhazmat.2020.123156

WANG, H. et al. Transcriptomic profiling sheds light on the blue-light and red-light response of oyster mushroom (Pleurotus ostreatus). Amb Express, v. 10, p. 1-10, 2020. https://doi.org/10.1186/s13568-020-0951-x

WIAFE-KWAGYAN, M.; ODAMTTEN, G.T.; KORTEI, N.K. Influence of substrate formulation on some morphometric characters and biological efficiency of Pleurotus ostreatus EM-1 (Ex. Fr) Kummer grown on rice wastes and "wawa" (Triplochiton scleroxylon) sawdust in Ghana. Food Science & Nutrition, v. 10, p. 1854-1863, 2022. https://doi.org/10.1002/fsn3.2802

YANG, Y.R.; GUO, Y.X.; WANG, Q.Y.; HU, B.Y.; TIAN, S.Y.; YANG, Q.Z.; CHENG, Z.A.; CHEN, Q.J.; ZHANG, G.Q. Impacts of composting duration on physicochemical properties and microbial communities during short-term composting for the substrate for oyster mushrooms. Sci. Total Environ. 2022, 847, 157673. https://doi.org/10.1016/j.scitotenv.2022.157673

YUE, Z. et al. Effect of Different Light Qualities and Intensities on the Yield and Quality of Facility-Grown Pleurotus eryngii. Journal of Fungi, v. 8, n. 12, p. 1244, 2022. https://doi.org/10.3390/jof8121244

ZAKIL, F. A. et al. Growth performance and mineral analysis of Pleurotus ostreatus (oyster mushroom) cultivated on spent mushroom medium mixed with rubber tree sawdust. Materials Today: Proceedings, v. 57, p. 1329-1337, 2022. https://doi.org/10.1016/j.matpr.2022.01.112

ZAKIL, F.A.; SUEB, M.S.M.; ISHA, R. Growth and yield performance of Pleurotus ostreatus on various agro-industrial wastes in Malaysia. Presented at the PROCEEDINGS OF THE 2ND INTERNATIONAL CONFERENCE ON BIOSCIENCES AND MEDICAL ENGINEERING (ICBME2019): Towards innovative research and cross-disciplinary collaborations, Bali, Indonesia, p. 020055, 2019. https://doi.org/10.1063/1.5125559

ZAMORA ZAMORA, Hernan Dario et al. Simultaneous production of cellulases, hemicellulases, and reducing sugars by Pleurotus ostreatus growth in one-pot solid state fermentation using Alstroemeria sp. waste. Biomass Conversion and Biorefinery, p. 1-14, 2021. https://doi.org/10.1007/s13399-021-01723-3

ZAWADZKA, A. et al. The effect of light conditions on the content of selected active ingredients in anatomical parts of the oyster mushroom (Pleurotus ostreatus L.). PLoS One, v. 17, n. 1, p. e0262279, 2022. https://doi.org/10.1371/journal.pone.0262279

ZHANG, W; LIU, S; Kuang, Y; ZHENG, S. Development of a novel spawn (block spawn) of an edible mushroom, Pleurotus ostreatus, in liquid culture and its cultivation evaluation. Mycobiology. Vol. 47, ed. 1, 2019. https://doi.org/10.1080/12298093.2018.1552648

ZHENG, Y. et al. Asexual reproduction and vegetative growth of Bionectria ochroleuca in response to temperature and photoperiod. Ecology and evolution, v. 11, n. 15, p. 10515-10525, 2021. https://doi.org/10.1002/ece3.7856

ZHOU, G.; PARAWIRA, W. The effect of different substrates found in Zimbabwe on the growth and yield of oyster mushroom Pleurotus ostreatus. Southern Africa Journal of Education, Science and Technology, v. 5, n. 2, p. 73-86, 2022. 10.4314/sajest.v5i2.39831

Printed by Books on Demand GmbH, Norderstedt / Germany